SpringerBriefs in Electrical and Computer Engineering

Control, Automation and Robotics

Series Editors

Tamer Başar
Antonio Bicchi
Miroslav Krstic

W0192899

For further volumes:
http://www.springer.com/series/10198

Zhuang Jiao · YangQuan Chen
Igor Podlubny

Distributed-Order Dynamic Systems

Stability, Simulation, Applications and Perspectives

 Springer

Dr. Zhuang Jiao
Department of Automation
Tsinghua University
Beijing
People's Republic of China

Prof. YangQuan Chen
Department of Electrical and Computer
    Engineering
Utah State University
Old Main Hill, CSOIS 4120
Logan, UT 84322-4120
USA

Prof. Igor Podlubny
BERG Faculty
Technical University of Kosice
B. Nemcovej 3
04200 Kosice
Slovakia

ISSN 2191-8112
ISBN 978-1-4471-2851-9
DOI 10.1007/978-1-4471-2852-6
Springer London Heidelberg New York Dordrecht

e-ISSN 2191-8120
e-ISBN 978-1-4471-2852-6

British Library Cataloguing in Publication Data
A catalogue record for this book is available from the British Library

Library of Congress Control Number: 2012931946

Printed on acid-free paper

Springer is part of Springer Science+Business Media (www.springer.com)

To our colleagues, friends, mentors and families

Preface

Fractional calculus is now being more widely accepted. The (constant) order of differentiation and/or integration can be an arbitrary real number including integers as special cases. For example, a low pass filter (LPF) with a fractional order pole can be written as $H(s) = 1/(\tau s^{\alpha_0} + 1)$, where $\alpha_0 > 0$ is a constant. Its corresponding governing differential equation is

$$\tau \frac{d^{\alpha_0}}{dt^{\alpha_0}} y(t) + y(t) = u(t),$$

where $u(t)$ and $y(t)$ are input and output signals, respectively; $\frac{d^{\alpha_0}}{dt^{\alpha_0}} y(t)$ or $y^{(\alpha_0)}(t)$ is the notation of fractional order derivative of $y(t)$. It is mathematically immediate to generalize this constant-order LPF in distributed-order sense as

$$H_{do}(s) = \frac{1}{b-a} \int_a^b \frac{1}{\tau s^{\alpha} + 1} d\alpha,$$

where a and b are given constants and the term $\frac{1}{b-a}$ is for scaling the DC gain to be 0 dB. The distributed-order dynamics can be characterized by the following distributed-order differential equation

$$\int_a^b w(\alpha) \frac{d^{\alpha}}{dt^{\alpha}} y(t) d\alpha + y(t) = u(t),$$

where $w(\alpha)$ can be regarded as order-dependent time constant or "order weight/distribution function."

Note that, the above constant-order model is in the same form of the famous classic Cole–Cole relaxation model, which can be recovered from the distributed-order model by setting the order distribution function $w(\alpha) = \delta(\alpha - \alpha_0)$, where $\delta(\cdot)$ is the well known Dirac Delta function. So, it is natural to believe that distributed-order Cole–Cole model $H_{do}(s)$ may be in a better position to characterize the *complex* material properties when the distribution function $w(\alpha)$ is properly chosen. The wisdom in modeling *"All models are wrong but some are useful"* and

"All models are wrong but some are dangerous", in fact, encourages us to explore the distributed-order generalization since we believe this notion is helpful, at least partially, as demonstrated in this Brief, with no harm.

With the above in mind, this Brief presents a general approach of distributed-order operator which can and will find its use for real world applications, as being observed from recent literature in many fields of science and engineering. It is devoted to provide an introduction of the latest research results about distributed-order dynamic system and control as well as distributed-order signal processing, which are based on the distributed-order differential/integral equations, to serve the control and signal processing community as a guide to understanding and using distributed-order differential/integral equations in order to enlarge the application domains of its disciplines, and to improve and generalize well established (constant-order) fractional-order control methods and strategies.

A major goal of this Brief is to present a concise and insightful view of the relevant knowledge by emphasizing fundamental methods and tools to understand why distributed-order concept is useful in control and signal processing, to understand its terminology, and to illuminate the key points of its applicability. The Brief is suitable for science and engineering community for broadening their toolbox in modeling, analysis, control, filtering tasks, with a hope that, transformative progress can be made in their respective research projects.

Tsinghua University   Zhuang Jiao
Utah State University   YangQuan Chen
Technical University of Kosice   Igor Podlubny
September 2011

Acknowledgements

As an emerging research topic, distributed-order systems started to draw attention from the research community. With this SpringerBriefs, we wish to provide an almost exhaustive, comprehensive literature review and a summary of our research efforts during the past few years on distributed-order systems. This Brief thus contains materials from papers and articles that were previously published. We are thankful and would like to acknowledge the copyright permissions from the following publishers who have released our work on that topic:

Acknowledgement is given to the Institute of Electrical and Electronic Engineers (IEEE) to reproduce material from the following papers:

©2010 IEEE. Reprinted, with permission, from Yan Li, Hu Sheng, and YangQuan Chen. "On distributed order low pass filter," *Proceedings of the 2010 IEEE/ASME International Conference on Mechatronic and Embedded Systems and Applications,"* Qingdao, ShanDong, China, 2010, pp. 588–592, doi:10.1109/MESA.2010.5552095 (some materials found in Chap. 4).

Acknowledgement is given to Elsevier B.V. to reproduce material from the following papers:

©2010 Elsevier B.V. Reprinted, with permission, from Yan Li, Hu Sheng, YangQuan Chen. "On distributed order integrator/differentiator", *Signal Processing*, vol. 91, no. 5, 2011, pp. 1079–1084, doi:10.1016/j.sigpro.2010.10.005 (some materials found in Chap. 4).

©2010 Elsevier B.V. Reprinted, with permission, from Igor Podlubny, Aleksei Chechkin, Tomas Skovranek, YangQuan Chen, Blas M. Vinagre Jara. "Matrix approach to discrete fractional calculus II: Partial fractional differential equations", *Journal of Computational Physics*, Volume 228, Issue 8, 1 May 2009, Pages 3137–3153. doi:10.1016/j.jcp.2009.01.014 (several figures and equations found in Chap. 5).

The research described in this Brief would not have been possible without the inspiration and help from the work of individuals in the research community, and we would like to acknowledge their help.

There are many people to whom the authors are obliged for their help and support.

Zhuang Jiao would like to express his sincere thanks to his tutor, Prof. Yisheng Zhong, the Tsinghua-Santander Postgraduate Research Scholarship for financial support to his research study in the USA, former and current CSOIS members: Dr. Yan Li, Dr. Hu Sheng for their support during his Ph.D. studies in CSOIS at Utah State University as an exchange Ph.D. visiting scholar. In particular, he appreciates the members of AFC (Applied Fractional Calculus) at Utah State University: Calvin Coopmans, Hadi Malek, Prof. Deshun Yin, Dr. Dali Chen, Dr. Haiyang Chao, Jinlu Han, Long Di, Yaojin Xu, Dr. Xuefeng Zhang, Dr. Kecai Cao, Peng Guo, Bo Li, Shuai Hu, Zhuo Li, Ms. Pooja Kavathekar, Ms. Sara Dadras.

YangQuan Chen would like to thank his wife Dr. Huifang Dou and his sons Duyun, David, and Daniel, for their patience, understanding, and complete support throughout this work. He is also thankful to Caibin Zeng for assistance in final round of proofreading.

Igor Podlubny is thankful to Aleksandr N. Vityuk and Viktor V. Verbitsky from the Odessa National University (Odessa, Ukraine) for fruitful discussions on distributed-order operators, and to Anatoly A. Alikhanov from the Kabardino-Balkarian State University (Nalchik, Russia), for providing help with obtaining some sources. Support from the BERG Faculty of the Technical University of Kosice (dean: Professor Gabriel Weiss) and from the grant agencies VEGA (grant 1/0497/11) and APVV (grants APVV-0040-07, APVV-0482-11, SK-UA-0042-09) is gratefully acknowledged.

Finally, we would like to thank Oliver Jackson of Springer for his interest in this Brief project and to Charlotte Cross, Editorial Assistant (Engineering), Springer London, for many good suggestions.

Contents

1 Introduction 1
  1.1 From Integer-Order Dynamic Systems to Fractional-Order
      Dynamic Systems. 1
  1.2 From Fractional-Order Dynamic Systems to Distributed-Order
      Dynamic Systems. 5
  1.3 Preview of Chapters 7
  1.4 Chapter Summary 8
  References 8

**2 Distributed-Order Linear Time-Invariant System (DOLTIS)
   and Its Stability Analysis** 11
  2.1 Introduction. 11
  2.2 Stability Analysis of DOLTIS in Four Cases. 11
  2.3 Time-Domain Analysis: Impulse Responses 19
  2.4 Frequency-Domain Response: Bode Plots 20
  2.5 Numerical Examples 21
  2.6 Chapter Summary 25
  References 28

**3 Noncommensurate Constant Orders as Special
   Cases of DOLTIS**. 29
  3.1 Introduction. 29
  3.2 Stability Analysis of Some Special Cases of DOLTIS 30
      3.2.1 Case 1: Double Noncommensurate Orders 30
      3.2.2 Case 2: N-Term Noncommensurate Orders 33
  3.3 Numerical Examples 34
  3.4 Chapter Summary 36
  References 36

4   Distributed-Order Filtering and Distributed-Order
Optimal Damping   39
   4.1   Application I: Distributed-Order Filtering   39
       4.1.1   Distributed-Order Integrator/Differentiator   39
       4.1.2   Distributed-Order Low-Pass Filter   45
       4.1.3   Impulse Response Invariant Discretization
            of DO-LPF   47
   4.2   Application II: Optimal Distributed-Order Damping.   49
       4.2.1   Distributed-Order Damping in Mass-Spring Viscoelastic
            Damper System   50
       4.2.2   Frequency-Domain Method Based Optimal
            Fractional-Order Damping Systems   52
   4.3   Chapter Summary   55
   References   56

5   Numerical Solution of Differential Equations
of Distributed Order   59
   5.1   Introduction.   59
   5.2   Triangular Strip Matrices   59
   5.3   Kronecker Matrix Product.   61
   5.4   Discretization of Ordinary Fractional Derivatives
       of Constant Order   62
   5.5   Discretization of Ordinary Derivatives of Distributed Order . . .   64
   5.6   Discretization of Partial Derivatives of Distributed Order   64
   5.7   Initial and Boundary Conditions for Using
       the Matrix Approach   67
   5.8   Implementation in MATLAB   67
   5.9   Numerical Examples   68
       5.9.1   Example 1: Distributed-Order Relaxation   69
       5.9.2   Example 2: Distributed-Order Oscillator.   70
       5.9.3   Example 3: Distributed-Order Diffusion   71
   5.10   Chapter Summary   72
   References   73

6   Future Topics   75
   6.1   Geometric Interpretation of Distributed-Order Differentiation
       as a Framework for Modeling   75
   6.2   From Positive Linear Time-Invariant Systems
       to Generalized Distributed-Order Systems.   76
   6.3   From PID Controllers to Distributed-Order PID Controllers . . .   78
   References   79

Appendix: MATLAB Codes   81

Index   89

Acronyms

BIBO	Bounded-Input Bounded-Output
CPE	Constant Phase Element
DOD	Distributed-order Differentiator
DODE	Distributed-order Differential Equation
DODS	Distributed-order Dynamic System
DOI	Distributed-order Integrator
DOIE	Distributed-order Integral Equation
DOLTIS	Distributed-order Linear Time-Invariant System
DOPDE	Distributed-order Partial Differential Equation
FC	Fractional Calculus
FODE	Fractional-order Differential Equation
FOIE	Fractional-order Integral Equation
FO-LTI	Fractional-order Linear Time-Invariant
FOPDE	Fractional-order Partial Differential Equation
IAE	Integral of Absolute Error
ISE	Integral of Squared Error
ISTE	Integral of Squared Time multiplied Error
ITAE	Integrated of Time multiplied Absolute Error
ITSE	Integrated of Time multiplied Squared Error
LPF	Low Pass Filter
NILT	Numerical Inverse Laplace Transform
ODE	Ordinary Differential Equation
PDE	Partial Differential Equation

Chapter 1
Introduction

1.1 From Integer-Order Dynamic Systems to Fractional-Order Dynamic Systems

As a branch of mathematics, calculus includes differential calculus and integral calculus. Calculus is the study of change, and has widespread applications in science, economics and engineering, and can solve many real world problems. It is well known that a system's dynamical properties can be described by an ordinary differential equation (ODE) which contains functions of an independent variable, and one or more of their derivatives with respect to that variable, for example, an ODE of the following form

$$F(x, y, y', \cdots, y^{(n-1)}, y^{(n)}) = 0$$

is called an ordinary differential equation of (integer) order n.

Being an important analytical tool in science and engineering, ordinary differential equation arises in many different fields including geometry, mechanics, astronomy and population modeling. Much attention has been devoted to the solution of ordinary differential equation. In the case where the equation is linear, it can be solved by analytical method; and there are several theorems that establish existence and uniqueness of solutions to initial value problems involving ordinary differential equations both locally and globally. Unfortunately, most of the interesting differential equations are non-linear and, with a few exceptions, can not be solved analytically exactly; approximate solutions can be obtained by using computer approximations (numerical ordinary differential equations).

Most of the discussions of control systems and controller design for control systems are usually based on models which are described by using ordinary differential equations. However, the physical quantity in many systems may depend on several independent variables. There is another type of differential equation when there are two or more independent variables, i.e., partial differential equation (PDE) of the following form

Z. Jiao et al., *Distributed-Order Dynamic Systems,* SpringerBriefs in Control, Automation and Robotics, DOI: 10.1007/978-1-4471-2852-6_1, © The Author(s) 2012

$$F\left(x_1, \cdots, x_n, u, \frac{\partial}{\partial x_1}u, \cdots, \frac{\partial}{\partial x_n}u, \frac{\partial^2}{\partial x_1 \partial x_1}u, \frac{\partial^2}{\partial x_1 \partial x_2}u, \cdots\right) = 0,$$

which involves partial derivatives of functions of several variables, and is a relation involving an unknown function (or functions) of several independent variables and their partial derivatives. Partial differential equations can be used to formulate, and thus aid the solution of, problems involving functions of several variables; such as the propagation of sound or heat, electrostatics, electrodynamics, fluid flow and elasticity. Just as ordinary differential equations often model dynamical systems, partial differential equations usually model multidimensional dynamical systems. There are several well known partial differential equations, for example,

- heat equation $u_t = \alpha u_{xx}$;
- wave equation $u_{tt} = c^2 u_{xx}$;
- Laplace equation $\varphi_{xx} + \varphi_{yy} = 0$;

and so on. In frequency domain, it is well known that the rational transfer functions of systems modeled by ordinary differential equations are called lumped-parameter dynamic systems; the irrational transfer functions of systems modeled by partial differential equations are called distributed-parameter systems.

However, all the orders in the above relevant ordinary differential equations and partial differential equations or the powers in the rational/irrational transfer functions are integers, curious researchers may have the question that why not the order be a rational, irrational, or even a complex number? This lead to the letter from Leibniz to L'Hospital at the very beginning of (integer-order) integral and differential calculus in 1695, in which Leibniz himself raised the question: "Can the meaning of derivatives with integer orders be generalized to derivatives with non-integer orders?" Until now the question raised by Leibniz for a non-integer-order derivative as an ongoing topic has been studied for more than 300 years, and it is known as fractional calculus (FC) (Miller and Ross 1993; Podlubny 1999) now, a generalization of calculus, which contains differentiation and integration of arbitrary (non-integer) order. However, it is necessary and important to make a clear statement that "fractional" or "fractional-order" is improperly used, a more accurate term should be "non-integer-order" since the order itself can be irrational, or complex number as well. The reason that we continue to use the term "fractional" is because a tremendous amount of work in the literatures use "fractional" more generally to refer to the same concept.

There are several well known definitions of fractional calculus operators, which are recalled in the following:

- Grünwald-Letnikov's fractional-order derivative/integral definition:

$$_a^G\mathrm{D}_t^\alpha f(t) := \lim_{h \to 0} \frac{1}{h^\alpha} \sum_{j=0}^{[(t-a)/h]} (-1)^j \binom{\alpha}{j} f(t - jh), \quad (\alpha \in \mathrm{R}).$$

- Riemann-Liouville's fractional-order integral definition:

$$_a^R D_t^{-\alpha} f(t) := \frac{1}{\Gamma(\alpha)} \int_a^t (t - \tau)^{\alpha - 1} f(\tau) d\tau, \ (\alpha > 0).$$

- Riemann-Liouville's fractional-order derivative definition:

$$_a^R D_t^\alpha f(t) := \frac{1}{\Gamma(n - \alpha)} \frac{d^n}{dt^n} \left[\int_a^t (t - \tau)^{n - \alpha - 1} f(\tau) d\tau \right], \ (n - 1 < \alpha < n).$$

- Caputo's fractional-order derivative definition:

$$_a^C D_t^\alpha f(t) := \frac{1}{\Gamma(n - \alpha)} \left[\int_a^t (t - \tau)^{n - \alpha - 1} f^{(n)}(\tau) d\tau \right], \ (n - 1 < \alpha < n).$$

Based on these definitions, the study on fractional calculus equations, i.e., fractional-order differential equation (FODE) and fractional-order integral equation (FOIE) which can describe more accurate behaviors of real physical phenomenon and systems have become a hot topic in the last decades. Fractional derivative provides a perfect tool when it is used to describe the memory and hereditary properties of various materials and processes, this is the main reason that fractional differential equations are being used in modeling mechanical and electrical properties of real materials, rheological properties of rocks, and many other fields. As an important application field of fractional calculus, the topic about fractional-order control and system has attracted many researchers to work on. A traditional fractional-order differential equation which can describe the fractional-order system's dynamical properties is of the following form:

$$F\left(x, {}_0D_t^{\alpha_1} y, {}_0D_t^{\alpha_2} y, \cdots, {}_0D_t^{\alpha_n} y\right) = 0$$

where ${}_0D_t^{\alpha_i}$, $(i = 1, \cdots, n)$ can adopt Riemann-Liouville's or Caputo's definition. Before discussing fractional-order systems and control, let us recall some traditional control concepts.

In feedback control, the basic control actions and their effects in the controlled system behavior are well known in the frequency domain. Note that these actions include proportional k, derivative s, and integral $1/s$, which are known as *PID* control, and their main effects over the controlled system behavior are Astrom and Murray (2008):

- for proportional action, it is to increase the speed of the response, and to decrease the steady-state error and relative stability;
- for derivative action, it is to increase the relative stability and the sensitivity to noise;

- for integral action, it is to eliminate the steady-state error, and to decrease the relative stability.

For the derivative action s, the positive effects (increased relative stability) can be observed in the frequency domain by the $\pi/2$ phase lead introduced, and the negative ones (increased sensitivity to high-frequency noise) by the increasing gain with slope of $20\,dB/dec$. The positive effects of integral action $1/s$ (elimination of steady-state errors) can be deduced by the infinite gain at zero frequency, and the negative ones (decreased relative stability) by the $\pi/2$ phase lag introduced. By considering the above, it is quite natural to have a conclusion that we could achieve more satisfactory compromises between the positive and negative effects by introducing more general control actions of the form s^α, $1/s^\beta$, with $\alpha, \beta > 0$, and we could develop more powerful and flexible design methods to satisfy the controlled system's specifications by combining the actions s^α and $1/s^\beta$. The terms s^α and $1/s^\beta$ are the essence of fractional-order PID ($PI^\lambda D^\mu$) control, and the traditional transfer function of fractional-order system is of the form

$$G(s) = \frac{b_1 s^{\beta_1} + b_2 s^{\beta_2} + \cdots + b_m s^{\beta_m}}{a_1 s^{\alpha_1} + a_2 s^{\alpha_2} + \cdots + a_n s^{\alpha_n}}.$$

Now let us focus our attention on system modeling. Researchers in viscoelasticity, electrochemistry, material science, biological systems and other fields in which diffusion, electrochemical, mass transport, or other memory phenomena appear (Bagley and Torvik 1984; Magin 2006), usually perform frequency domain experiments in order to obtain the equivalent electrical circuits which can reflect the same dynamic behaviors of the actual systems. It is quite normal in these fields to find behaviors that are not the expected ones for common lumped elements (resistors, inductors and capacitors) at all, and to define some special impedances such as constant phase elements (CPEs), Warburg impedances, and others for operational purposes. All these proposed special impedances have in common the frequency domain responses of the form $k/(j\omega)^\alpha$, $\alpha \in R$, and should be modelled by k/s^α, $\alpha \in R$ in the Laplace domain. These operators mentioned above can lead to the corresponding operators in the time domain, which are the definitions of differential and integral operators of arbitrary order, i.e., the fundamental operators of the fractional calculus. Similar to the relationship between ordinary differential equations and fractional-order differential equations, there are fractional-order partial differential equations corresponding to the partial differential equations. The well known fractional-order partial differential equations (FOPDE) are recalled as following:

- Time fractional-order diffusion equation:

$$\frac{\partial^\alpha u(x,t)}{\partial t^\alpha} = \frac{\partial^2 u(x,t)}{\partial x^2}, \quad (0 < \alpha \le 1).$$

- Time fractional-order wave equation:

$$\frac{\partial^{\alpha} u(x,t)}{\partial t^{\alpha}} = \frac{\partial^2 u(x,t)}{\partial x^2}, \quad (1 < \alpha \le 2).$$

- Time fractional-order diffusion-wave equation:

$$a\frac{\partial^{\alpha} u(x,t)}{\partial t^{\alpha}} + b\frac{\partial^{\beta} u(x,t)}{\partial t^{\beta}} = \frac{\partial^2 u(x,t)}{\partial x^2}, \quad (0 < \alpha \le 1 < \beta \le 2).$$

As an interdisciplinary branch of fractional calculus and control engineering, the system can be modeled in a classical way or as a fractional-order one; the controller can also be operated as a classical one or a fractional-order one. Then there are four strategies of control systems, which are integer-order controller for integer-order system, integer-order controller for fractional-order system, fractional-order controller for integer-order system and fractional-order controller for fractional-order system. In the last several decades, there has been continuing growth of papers discussing the issues of fractional-order systems and controls, for example, the stability results on fractional-order linear time-invariant (or FOLTI) systems with commensurate orders were presented in Matignon (1996) for the first time; $PI^{\lambda}D^{\mu}$ controller, a generalization of *PID* controller was proposed in Podlubny (1999); the tuning rule and experiments of fractional order proportional and derivative (FOPD) motion controller were given in Li et al. (2009); CRONE Control (Oustaloup et al. 1995) was the first robust control method based on fractional differentiation for linear time-invariant systems; the systematic results on the robust stability of interval uncertain FOLTI systems were presented in Ahn et al. (2007), Ahn and Chen (2008), Chen et al. (2006), Lu and Chen (2009, 2010); the bounded-input bounded-output (BIBO) stability of fractional-order delay systems of retarded and neutral types was studied in Bonnet and Partington (2002, 2007); based on Cauchy's integral theorem and by solving an initial-value problem, an effective numerical algorithm for testing the BIBO stability of fractional delay systems was presented in Hwang and Cheng (2006). The latest monograph (Caponetto et al. 2010; Lu and Chen 2009) gave the systematic knowledge about fractional-order dynamic systems and controls.

1.2 From Fractional-Order Dynamic Systems to Distributed-Order Dynamic Systems

When the fractional calculus operators act on $f(t)$, and we integrate $_0D_t^{\alpha} f(t)$ with respect to the order, then distributed-order differential/integral equations can be obtained. In this Brief, the following distributed-order differential/integral operator notation is adopted:

$$_0\mathrm{D}_t^{w(\alpha)} f(t) := \int_{\gamma_1}^{\gamma_2} w(\alpha)_0\mathrm{D}_t^\alpha f(t)\mathrm{d}\alpha$$

where $w(\alpha)$ denotes the weight function of distribution of order $\alpha \in [\gamma_1, \gamma_2]$.

The idea of distributed-order equation was first proposed by Caputo (1969) and solved by him in 1995 (Caputo 1995). The distributed-order equation is intended to model the input–output relationship of a linear time-invariant system based on the frequency domain response observation, i.e., the distributed-order equation is the time domain representation of the input–output relationship observed and constructed in the frequency domain. The general form of the distributed-order differential equation (DODE) can be given as:

$$\sum_{i=1}^N a_i \int_0^1 w_i(\alpha)_0\mathrm{D}_t^{i-\alpha} x(t)\mathrm{d}\alpha + \sum_{j=0}^N b_j x^{(j)}(t) = f(t)$$

where $w_i(\alpha)$ denotes the weight function with respect to the order $\alpha \in [\gamma_1, \gamma_2]$. From now on, we note that the above equation can be viewed as the generalization of ordinary differential equation ($w_i(\alpha) \equiv 0$) or fractional-order differential equation ($w_i(\alpha)$ takes only discrete values in $[\gamma_1, \gamma_2]$), then it can be concluded that both integer-order systems and fractional-order systems are special cases of distributed-order systems (Lorenzo and Hartley 2002).

Recently, much attention has been paid to the distributed-order differential equations and their applications in engineering fields. For example, the general solution of linear distributed-order differential equation was discussed systematically in Bagley and Torvik (2000); distributed-order equations were introduced in the constitutive equations of dielectric media (Caputo 1995), the distributed-order fractional kinetics was discussed in Sokolov et al. (2004); the multi-dimensional random walk models were governed by distributed fractional order differential equations in Umarov and Steinberg (2006); particularly, the distributed-order operator becomes a more precise tool to explain and describe some real physical phenomena such as the complexity of nonlinear systems (Adams et al. 2008; Atanackovic et al. 2007, 2009b, c; Diethelm and Ford 2009; Hartley and Lorenzo 2003; Lorenzo and Hartley 1998, 2002; Mainardi et al. 2007a; Sokolov et al. 2004), networked structures (Carlson and Halijak 1964; Lorenzo and Hartley 2002; Xu and Tan 2006), nonhomogeneous phenomena (Caputo 2001; Chen et al. 2009; Kochubei 2008; Srokowski 2008; Sun et al. 2009, 2010; Umarov and Steinberg 2006), multi-scale and multi-spectral phenomena (Atanackovic et al. 2005; Bohannan 2000; Connolly 2004; Mainardi et al. 2008; Mainardi and Pagnini 2007; Tsao 1987), etc.

Besides the distributed-order differential equations, there are still distributed-order partial differential equations (DOPDE) being studied as the following:

$$_0\mathrm{D}_t^{w(\alpha)} u(x, t) = \frac{\partial^2}{\partial x^2} u(x, t)$$

where $w(\alpha)$ denotes the function of distribution of order $\alpha \in [0, 2]$.

Let supp denotes the support set, and we can set $\alpha_1 := \inf \{\alpha \,|\, \alpha \in \text{supp} \, w(\alpha)\}$, and $\alpha_2 := \sup \{\alpha \,|\, \alpha \in \text{supp} \, w(\alpha)\}$, then the following cases can be distinguished:

- Time distributed-order diffusion-wave equation: $0 < \alpha_1 \leq 1 < \alpha_2 \leq 2$;
- Time distributed-order diffusion equation : $\alpha_2 \leq 1$;
- Time distributed-order wave equation : $\alpha_1 > 1$.

For the distributed-order partial differential equations, there have been some papers discussing those problems. For example, the time distributed-order diffusion-wave equation was considered in Atanackovic et al. (2009b, c); time-fractional diffusion of distributed order was discussed in Mainardi et al. (2007b, 2008); distributed-order wave equation was analyzed in Atanackovic et al. (2011), for more knowledge about distributed-order partial differential equations, please refer to Atanackovic et al. (2009a), Chechkin et al. (2002), Mainardi and Pagnini (2007), Meerschaert et al. (2011).

The theories of the distributed-order equations can be classified as: distributed-order equations (Atanackovic et al. 2009b, c; Bagley and Torvik 2000; Caputo 1995), distributed-order system identification (Hartley and Lorenzo 2003; Sokolov et al. 2004; Srokowski 2008), special functions in distributed-order calculus (Atanackovic et al. 2009a; Caputo 2001; Mainardi et al. 2007a; Mainardi and Pagnini 2007), numerical methods (Chen et al. 2009; Diethelm and Ford 2009; Sun et al. 2009, 2010) and so on (Atanackovic et al. 2005; Kochubei 2008). Moreover, there are also three surveys (Lorenzo and Hartley 1998, 2002; Umarov and Steinberg 2006) and three thesis (Bohannan 2000; Connolly 2004; Tsao 1987) discussing the theories and applications of distributed-order operators. It is noted that the time domain analysis of the distributed order operator is still unmature and urgently needed to be developed. So in this Brief, some latest results are given in Chap. 5 with several worked out examples with MATLAB codes given in the appendices.

1.3 Preview of Chapters

In this chapter, we focus on setting up a concise context of our Brief theme by presenting our thought on progressing from integer-order system to fractional-order system, from fractional-order system to distributed-order system.

Chapter 2 is dedicated to the stability issue of distributed-order linear time-invariant (LTI) systems. Four different order distribution functions are analyzed in details. This chapter offers original and fundamental stability results for LTI distributed-order dynamic systems (DODS). Graphical and numerical results are included to show the fundamental differences compared to constant order and integer order LTI dynamic systems.

Chapter 3 serves the purpose of showing that DODS is a generalized model which is so powerful that some really hard research problems like stability of noncommensurate order LTI systems can be readily answered. Specifically, as the special cases of distributed-order linear time-invariant systems, the stability analysis of fractional-order systems with double noncommensurate orders and N-term noncommensurate orders are studied in Chap. 3.

Chapter 4 shows two generic application examples using distributed-order operator: distributed-order signal processing and optimal distributed-order damping. In distributed-order signal processing, the simplest case of distributed-order integrator/differentiator is discussed first followed by the discussion of distributed-order low-pass filter. Then, optimal distributed-order damping strategies are given for a given standard form of second order system knows as distributed-order fractional mass-spring viscoelastic damper system. Frequency-domain method based optimal fractional-order damping systems are numerically solved.

In Chap. 5, a new general approach to discretization of distributed-order derivatives and integrals and to numerical solution of ordinary and partial differential equations of distributed order is presented.

In Chap. 6, future topics related to distributed-order operator are discussed.

More than 100 reference are listed and cited in this Brief, even if it can not be a complete bibliography for this field of interest. Readers can find other reference related to this emerging topic.

MATLAB codes are provided as appendices so that the presented results of this Brief are reproducible, minimizing the repetitive coding work for beginners who decide to dive into this exciting and promising field of basic and applied research which is full of opportunities of transformative research. Anyway, distributed-order operator, is in fact characterizing mixed-scale dynamics, or trans-scale, or cross-scale dynamics as we see it.

1.4 Chapter Summary

In this chapter we have introduced the progression from integer-order dynamic systems to fractional-order dynamic systems, and from fractional-order dynamic systems to distributed-order dynamic systems. Basic notations together with literature reviews are presented. A brief chapter preview is included as well.

References

Adams JL, Hartley TT, Lorenzo CF (2008) Identification of complex order-distributions. J Vib Control 14(9–10):1375–1388

Ahn HS, Chen YQ (2008) Necessary and sufficient stability condition of fractional-order interval linear systems. Automatica 44(11):2985–2988

Ahn HS, Chen YQ, Podlubny I (2007) Robust stability test of a class of linear time-invariant interval fractional-order systems using Lyapunov inequality. Appl Math Comput 187(1):27–34

Astrom KJ, Murray RM (2008) Feedback systems: an introduction for scientists and engineers. Princeton University Press, Princeton

Atanackovic TM, Budincevic M, Pilipovic S (2005) On a fractional distributed-order oscillator. J Phys A: Math Gen 38(30):6703–6713

Atanackovic TM, Oparnica L, Pilipovic S (2007) On a nonlinear distributed order fractional differential equation. J Math Anal Appl 328(1):590–608

Atanackovic TM, Pilipovic S, Zorica D (2009a) Existence and calculation of the solution to the time distributed order diffusion equation. Phys Scripta T136:014012 (6pp)

Atanackovic TM, Pilipovic S, Zorica D (2009b) Time distributed-order diffusion-wave equation. I. Volterra-type equation. Proc Royal Soc A 465:1869–1891

Atanackovic TM, Pilipovic S, Zorica D (2009c) Time distributed-order diffusion-wave equation. II. Applications of Laplace and Fourier transformations. Proc Royal Soc A 465:1893–1917

Atanackovic TM, Pilipovic S, Zorica D (2011) Distributed-order fractional wave equation on a finite domain stress relaxation in a rod. Int J Eng Sci 49(2):175–190

Bagley RL, Torvik PJ (1984) On the appearance of the fractional derivative in the behavior of real materials. ASME J Appl Mech 51(2):294–298

Bagley RL, Torvik PJ (2000) On the existence of the order domain and the solution of distributed order equations (Parts I, II). Int J Appl Mech 2(7):865–882, 965–987

Bohannan G (2000) Application of fractional calculus to polarization dynamics in solid dielectric materials. PhD Dissertation, Montana State University, Nov 2000

Bonnet C, Partington JR (2002) Analysis of fractional delay systems of retarded and neutral type. Automatica 38(7):1133–1138

Bonnet C, Partington JR (2007) Stabilization of some fractional delay systems of neutral type. Automatica 43(12):2047–2053

Caponetto R, Dongola G, Fortuna L, Petras I (2010) Fractional order systems: modeling and control applications. World Scientific Company, Singapore

Caputo M (1969) Elasticità e dissipazione. Zanichelli, Bologna

Caputo M (1995) Mean fractional-order-derivatives differential equations and filters. Annali dell'Universita di Ferrara 41(1):73–84

Caputo M (2001) Distributed order differential equations modelling dielectric induction and diffusion. Fract Calc Appl Anal 4(4):421–442

Carlson G, Halijak C (1964) Approximation of fractional capacitors $(1/s)^{(1/n)}$ by a regular Newton process. IEEE Trans Circuit Theory 11(2):210–213

Chechkin AV, Gorenflo R, Sokolov IM (2002) Retarding subdiffusion and accelerating superdiffusion governed by distributed-order fractional diffusion equations. Phys Rev E 66:046129

Chen YQ, Ahn HS, Podlubny I (2006) Robust stability check of fractional order linear time invariant systems with interval uncertainties. Signal Process 86(10):2611–2618

Chen W, Sun HG, Zhang XD, Korosak D (2009) Anomalous diffusion modeling by fractal and fractional derivatives. Comput Math Appl 59(5):1754–1758

Connolly JA (2004) The numerical solution of fractional and distributed order differential equations. Thesis, University of Liverpool, Dec 2004

Diethelm K, Ford NJ (2009) Numerical analysis for distributed-order differential equations. J Comput Appl Math 225(1):96–104

Hartley TT, Lorenzo CF (2003) Fractional-order system identification based on continuous order-distributions. Signal Process 83(11):2287–2300

Hwang C, Cheng YC (2006) A numerical algorithm for stability testing of fractional delay systems. Automatica 42(5):825–831

Kochubei AN (2008) Distributed order calculus and equations of ultraslow diffusion. J Math Anal Appl 340(1):252–281

Li HS, Luo Y, Chen YQ (2009) A fractional order proportional and derivative (fopd) motion controller: tuning rule and experiments. IEEE Trans Control Syst Technol 18(2):1–5

Lorenzo CF, Hartley TT (1998) Initialization, conceptualization, and application in the generalized fractional calculus. NASA technical paper, NASA/TP 1998-208415

Lorenzo CF, Hartley TT (2002) Variable order and distributed order fractional operators. Nonlinear Dyn 29(1–4):57–98

Lu JG, Chen GR (2009) Robust stability and stabilization of fractional-order interval systems: an lmi approach. IEEE Trans Autom Control 54(6):1294–1299

Lu JG, Chen YQ (2010) Robust stability and stabilization of fractional order interval systems with the fractional order α: The $0 < \alpha < 1$ case. IEEE Trans Autom Control 55(1):152–158

Magin RL (2006) Fractional calculus in bioengineering. Begell House, Connecticut

Mainardi F, Pagnini G (2007) The role of the fox-wright functions in fractional sub-diffusion of distributed order. J Comput Appl Math 207(2):245–257

Mainardi F, Mura A, Gorenflo R, Stojanovic M (2007a) The two forms of fractional relaxation of distributed order. J Vib Control 9:1249–1268

Mainardi F, Mura A, Pagnini G, Gorenflo R (2007b) Some aspects of fractional diffusion equations of single and distributed order. Appl Math Comput 187:295–305

Mainardi F, Mura A, Pagnini G, Gorenflo R (2008) Time-fractional diffusion of distributed order. J Vib Control 14(9–10):1267–1290

Matignon D (1996) Stability results on fractional differential equations with applications to control processing. In: Multiconference on computational engineering in systems and application, pp 963–968

Meerschaert MM, Nane E, Vellaisamy P (2011) Distributed-order fractional diffusions on bounded domains. J Math Anal Appl 379:216–228

Miller KS, Ross B (1993) An introduction to the fractional calculus and fractional differential equations. Wiley, New York

Oustaloup A, Mathieu B, Lanusse P (1995) The crone control of resonant plants: application to a flexible transmission. Eur J Control 1(2):113–121

Podlubny I (1999) Fractional differential equations. Academic Press, San Diego

Podlubny I (1999) Fractional-order systems and $PI^\lambda D^\mu$ controllers. IEEE Trans Autom Control 44(1):208–214

Sokolov IM, Chechkin AV, Klafter J (2004) Distributed-order fractional kinetics. Acta Phys Polonica B 35(4):1323

Srokowski T (2008) Lévy flights in nonhomogeneous media: distributed-order fractional equation approach. Phys Rev E 78(3):031135

Sun HG, Chen W, Chen YQ (2009) Variable-order fractional differential operators in anomalous diffusion modeling. Phys A: Stat Mech Appl 388(21):4586–4592

Sun HG, Chen W, Sheng H, Chen YQ (2010) On mean square displacement behaviors of anomalous diffusions with variable and random orders. Phys Lett A 374(7):906–910

Tsao YY (1987) Fractal concepts in the analysis of dispersion or relaxation processes. PhD Dissertation, Drexel University, June 1987

Umarov S, Steinberg S (2006) Random walk models associated with distributed fractional order differential equations. Inst Math Stat 51:117–127

Xu MY, Tan WC (2006) Intermediate processes and critical phenomena: theory, method and progress of fractional operators and their applications to modern mechanics. Sci China: Ser G Phys Mech Astron 49(3):257–272

Chapter 2
Distributed-Order Linear Time-Invariant System (DOLTIS) and Its Stability Analysis

2.1 Introduction

By using distributed-order concept, we can describe the dynamical properties of real world system more accurately, so distributed-order system identification problem was studied in Hartley and Lorenzo (1999, 2003, 2004). In the following sections, the stability analysis of distributed-order linear time-invariant systems in four cases are first studied, then the frequency-domain responses are presented, and time-domain responses on the basis of numerical inverse Laplace transform technique are shown in details.

2.2 Stability Analysis of DOLTIS in Four Cases

Consider a distributed-order system described by the following linear time-invariant (LTI) distributed-order differential equation (DODE) and algebraic output equation:

$$_0D_t^{w(\alpha)}x(t) = \int_0^1 w(\alpha)_0D_t^\alpha x(t)d\alpha = Ax(t) + Bu(t)$$
$$y(t) = Cx(t) + Du(t) \tag{2.1}$$

where $w(\alpha)$ is the function of distribution of order $\alpha \in [0, 1]$, $_0D_t^\alpha$ denotes the Caputo fractional-order derivative operator, A, B, C, D are matrices with appropriate dimensions.

Remark 2.1 Since any interval (γ_1, γ_2) can be converted to $(0, 1)$ through variable substitution, without loss of generality, the integral interval in (2.1) is considered to be $(0, 1)$.

Z. Jiao et al., *Distributed-Order Dynamic Systems,* SpringerBriefs in Control, Automation and Robotics, DOI: 10.1007/978-1-4471-2852-6_2,
© The Author(s) 2012

For the distributed-order derivative operator $D^{w(\alpha)}x(t)$, the Laplace transform is

$$L\left\{D^{w(\alpha)}x(t)\right\}(s) = \tilde{x}(s)\int_0^1 w(\alpha)s^\alpha d\alpha - x(0)\frac{1}{s}\int_0^1 w(\alpha)s^\alpha d\alpha, \quad s \in \mathbb{C}\setminus(-\infty, 0]$$

where $\tilde{x}(s) = L\{x(t)\}(s) := \int_0^\infty x(t)e^{-st}dt$. By applying the Laplace transform to (2.1) with the assumptions that $x(0) = 0$, $u(t) = \delta(t)$ ($\delta(t)$ is the Dirac delta distribution), one obtains

$$\tilde{x}(s)\int_0^1 w(\alpha)s^\alpha d\alpha = A\tilde{x}(s) + B$$

i.e.,

$$\tilde{x}(s) = \left(\left(\int_0^1 w(\alpha)s^\alpha d\alpha\right)I - A\right)^{-1}B$$

where I is the identity matrix. Application of the inverse Laplace transform to the previous expression yields

$$x(t) = L^{-1}\left[\left(\left(\int_0^1 w(\alpha)s^\alpha d\alpha\right)I - A\right)^{-1}B\right](t), \quad t > 0. \tag{2.2}$$

In the following, four different cases of the weighting function of order are discussed respectively.

Case 1 $w(\alpha) = 1$

In this case, it can be followed by (2.2) that

$$\begin{aligned}
x(t) &= L^{-1}\left[\left(\left(\int_0^1 s^\alpha d\alpha\right)I - A\right)^{-1}B\right](t)\\
&= L^{-1}\left[\left(\frac{s-1}{\ln s}I - A\right)^{-1}B\right](t)\\
&= L^{-1}\left[\ln s(sI - (I + \ln sA))^{-1}B\right](t). \tag{2.3}
\end{aligned}$$

Remark 2.2 It is well known from complex analysis (Asmar and Jones 2002) that complex logarithm $\ln z = \ln |z| + i \arg z$ $(z \neq 0)$ defines a multiple-valued function, because $\arg z$ is multiple-valued. For term $\ln s$ in (2.3), we know that it is a multi-valued function of the complex variable s whose domain can be seen as a Riemann surface (Cuadrado and Cabanes 1989; Westerlund and Ekstam 1994) of a number of sheets which is infinite. Note that in multiple-valued functions only the first Riemann sheet has its physical significance (Gross and Braga 1961), so we can make $\ln s$ a single-valued function by specifying a single-valued $-\pi < \arg s < \pi$. Because

$s = 0$ and s on the negative real axis are nonremovable discontinuities, the branch cut of $\ln s$ is $(-\infty, 0]$.

Definition 2.1 A distributed-order system $H(s)$ defined by its impulse response $h(t) = L^{-1}\{H(s)\}$ is BIBO stable if and only if $\forall u \in L^{\infty}(R^+)$, $h * u \in L^{\infty}(R^+)$. $*$ stands for the convolution product and $L^{\infty}(R^+)$ stands for the Lebesgue space of measurable function h such that $ess \sup\limits_{t \in R^+} |h(t)| < \infty$.

Based on Definition 2.1 and the above analysis, the following theorem can be established.

Theorem 2.1 *The distributed-order linear time-invariant system (2.1) with transfer function $G_1(s) = C \ln s(sI - (I + A \ln s))^{-1} B + D$ is BIBO stable, if and only if all the eigenvalues of A lie on the left of curve $l_1 := l_a \bigcup l_b$ in the complex plane, where l_a and l_b are symmetrical with respect to the real axis, and*

$$l_a := \left\{ x - iy \,\middle|\, x = \frac{2\pi\omega - 4\ln\omega}{4(\ln\omega)^2 + \pi^2}, y = \frac{4\omega\ln\omega + 2\pi}{4(\ln\omega)^2 + \pi^2} \right\}$$

with $\omega \in [0, \infty)$.

Proof (if part) Note that the final value theorem implies that $\lim\limits_{t \to \infty} g(t) = sG_1(s) \to 0$, if all poles of $sG_1(s)$ are in the left half-plane when $s \to 0$. It can be easily known that all the poles of $sG_1(s)$ satisfy the transcendental characteristic equation of the form

$$|(s - 1)I - A \ln s| = 0. \tag{2.4}$$

From (2.4) we know that $\frac{s-1}{\ln s} = \sigma_i(A)$ $(i = 1, \cdots, n)$, where $\sigma(A)$ denotes the set of eigenvalues of A. As all the zeros of (2.4) should lie in the left half-plane to ensure the BIBO stability of distributed-order system $G_1(s)$, it is necessary to derive the range of $\lambda = \frac{s-1}{\ln s}$ when s belongs to the left half-plane.

It is natural to determine the range of $\lambda = \frac{s-1}{\ln s}$ when s lies on the imaginary axis. Then, for $s = j\omega$, $(-\infty < \omega < 0)$, we have

$$\lambda = \frac{(2\pi(-\omega) - 4\ln(-\omega)) + j(4(-\omega)\ln(-\omega) + 2\pi)}{4(\ln(-\omega))^2 + \pi^2}$$

while for $s = j\omega$, $(0 \leq \omega < \infty)$, we have

$$\lambda = \frac{(2\pi\omega - 4\ln\omega) - j(4\omega\ln\omega + 2\pi)}{4(\ln\omega)^2 + \pi^2}$$

which means that the imaginary axis is mapped to a curve denoted by l_1, which is symmetrical with respect to the real axis. By choosing a point s randomly which lies on the left of the imaginary axis, the range of $\lambda = \frac{s-1}{\ln s}$ lies on the left of curve l_1, which means that the stable region of distributed-order system (2.1) is the left region

Fig. 2.1 The stable boundary
of the distributed-order system
(2.1) $G_1(s)$

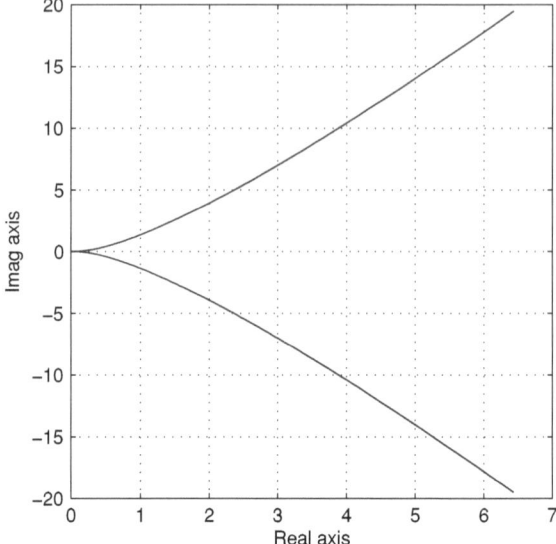

Fig. 2.2 The stable boundary
of the distributed-order system
(2.1) $G_1(s)$ (*Zoomed*)

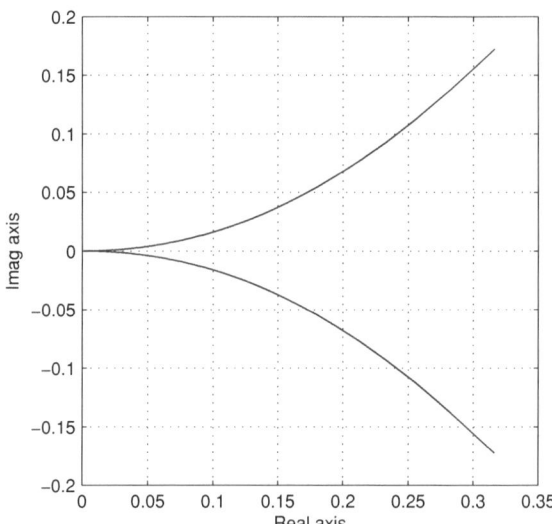

of curve l_1. In the following, l_1 is plotted in Fig. 2.1, with the local property around 0 zoomed in Fig. 2.2.

It can be easily known from the above analysis that if all the eigenvalues of A lie on the left of curve l_1, all the poles of $sG_1(s)$ lie on the left half-plane. From the final value theorem, we further know that $\lim_{t \to \infty} g(t) = 0$, which means the distributed-order system with transfer function $G_1(s) = C \ln s(sI - (I + A \ln s))^{-1}$ $B + D$ is BIBO stable.

(only if part) It is obviously known from Definition 2.1 that $G_1(s)$ lies in H_∞, the space of bounded analytic functions on the right half plane of the complex plane, which means that all the poles of $G_1(s)$ lie in the left half plane of the complex plane. From the proof of (if part), it is known that $\{s_k\}_{k=1,2,\cdots,n}$ lie in the open left half plane, which is equivalents to that all the eigenvalues of A lie in the left region with respect to l_1.

Remark 2.3 It is easy to conclude that the slope of the curve l_1 at the original point is 0, and is infinity at the infinite point, which means that any ray in the first quadrant starts at point 0 will have point of intersection with the curve l_1. This means any constant fractional-order approximation of DODS is problematic, since the stability domains are different.

Case 2 $w(\alpha) = \alpha$

In this case, the following can be obtained under the similar analysis procedure in Case 1,

$$x(t) = L^{-1}\left[\left(\left(\int_0^1 \alpha s^\alpha d\alpha\right)I - A\right)^{-1}B\right]$$

$$= L^{-1}\left[\left(\frac{1 - s + s \ln s}{\ln^2 s}I - A\right)^{-1}B\right]$$

$$= L^{-1}\left[\ln^2 s\left((1 - s + s \ln s)I - \ln^2 s A\right)^{-1}B\right].$$

Theorem 2.2 *The distributed-order linear time-invariant system (2.1) with transfer function* $G_2(s) = C\ln^2 s((1 - s + s \ln s)I - A\ln^2 s)^{-1}B + D$ *is BIBO stable, if and only if all the eigenvalues of A lie on the left of curve* $l_2 := l_c \bigcup l_d$, *where l_c and l_d are symmetrical with respect to the real axis, and* $l_c := \{x + iy \,|\, x = x_\omega, y = y_\omega, \omega \in (0, \infty)\}$, *with notations*

$$x_\omega = \frac{\left(\ln \omega - \frac{\pi^2}{4}\right)\left(1 - \frac{\pi}{2}\omega\right) - \pi \ln \omega \left(\omega - \omega \ln \omega\right)}{\left(\ln^2 \omega + \frac{\pi^2}{4}\right)^2}$$

and

$$y_\omega = \frac{\left(\ln \omega - \frac{\pi^2}{4}\right)\left(\omega - \omega \ln \omega\right) + \pi \ln \omega \left(1 - \frac{\pi}{2}\omega\right)}{\left(\ln^2 \omega + \frac{\pi^2}{4}\right)^2}.$$

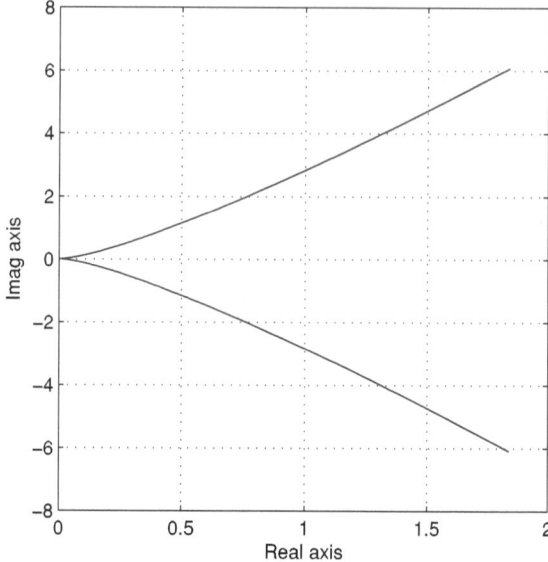

Fig. 2.3 The stable boundary of distributed-order system (2.1) $G_2(s)$

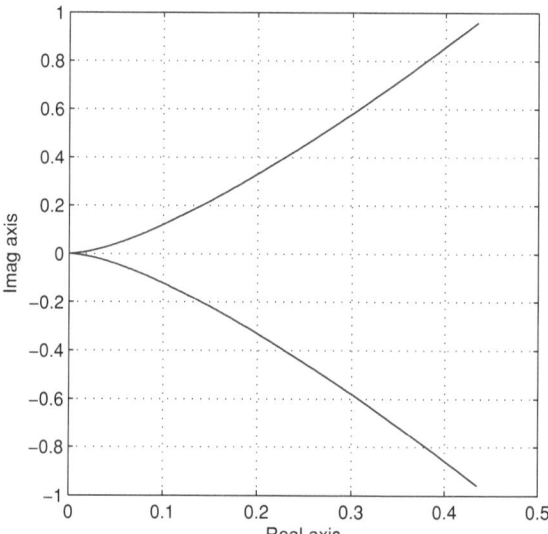

Fig. 2.4 The stable boundary of distributed-order system (2.1) $G_2(s)$ (*Zoomed*)

The proof of Theorem 2.2 can be given by the similar procedures in Theorem 2.1, the stable boundary for distributed-order system $G_2(s)$ is shown in Fig. 2.3, with the local property around 0 shown in Fig. 2.4.

Case 3 $w(\alpha) = \delta(\alpha - \beta), \ (0 < \beta < 1)$

In this case, the DODE (2.1) converts to a constant-order fractional-order system described by

$$\begin{aligned} {}_0D_t^\beta x(t) &= Ax(t) + Bu(t) \\ y(t) &= Cx(t) + Du(t). \end{aligned} \tag{2.5}$$

Using Laplace transform, the irrational transfer function of fractional-order system (2.5) with null initial conditions is

$$G_3(s) = C\left(s^\beta I - A\right)^{-1} B + D. \tag{2.6}$$

Remark 2.4 Note that term s^β in (2.6) defines a multi-valued function of the complex variable s whose domain can be seen as a Riemann surface (Cuadrado and Cabanes 1989; Westerlund and Ekstam 1994) of a number of sheets which is finite in the case of $\beta \in Q^+$, and infinite in the case of $\beta \in R^+ \backslash Q^+$. It is well known that in multiple-valued functions only the principal sheet defined by $-\pi < \arg s < \pi$ has its physical significance (Gross and Braga 1961).

The following can be obtained under the similar analysis procedure in the previous cases,

$$\begin{aligned} x(t) &= L^{-1}\left[\left(\left(\int_0^1 \delta(\alpha - \beta)s^\alpha d\alpha\right) I - A\right)^{-1} B\right](t) \\ &= L^{-1}\left[\left(s^\beta I - A\right)^{-1} B\right](t). \end{aligned}$$

The following theorem which corresponds to the stability condition of fractional-order system obtained in Matignon (1996) can be given.

Theorem 2.3 *The fractional-order linear time-invariant system with transfer function* $G_3(s) = C\left(s^\beta I - A\right)^{-1} B + D$ *is BIBO stable, if and only if all the eigenvalues of A lie on the left of curve* $l_3 := l_e \bigcup l_f$, *where* l_e *and* l_f *are symmetrical with respect to the real axis, and* $l_e := \left\{ re^{i\theta} \mid r = \omega^\beta, \theta = \pi\beta/2, \omega \in (0, \infty) \right\}$.

The proof of Theorem 2.3 can be given by the similar procedures in Theorem 2.1, the stable region for fractional-order system $G_3(s)$ with $\beta = 0.5$ is shown in Fig. 2.5.

Case 4 $w(\alpha) = \sum_{k=1}^n b_k \delta(\alpha - k\beta), \ (0 < n\beta < 1)$.

In this case, the DODE (2.1) converts to the so-called LTI commensurate fractional-order system

Fig. 2.5 The stable boundary
of fractional-order system
(2.5) $G_3(s)$

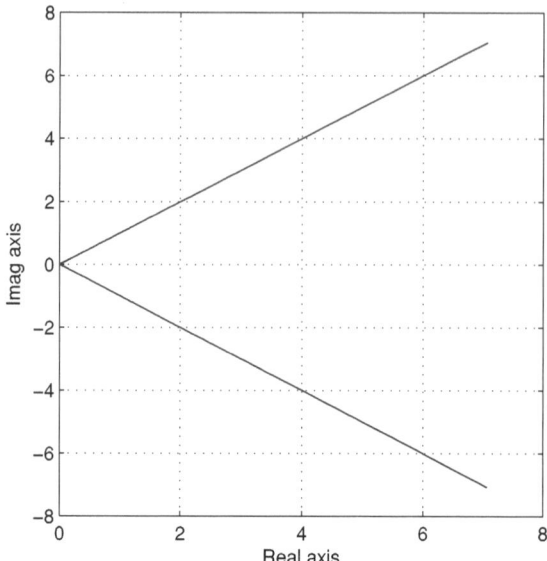

$$\sum_{k=1}^{n} b_{k0} D_t^{k\beta} x(t) = Ax(t) + Bu(t)$$

$$y(t) = Cx(t) + Du(t). \tag{2.7}$$

Let $\hat{x}(t) = \begin{bmatrix} x(t) & D^\beta x(t) & D^{2\beta} x(t) \cdots D^{(n-1)\beta} x(t) \end{bmatrix}^T$, (2.7) can be converted to the following equivalent form

$$_0 Dt^\beta \hat{x}(t) = \hat{A}\hat{x}(t) + \hat{B}u(t) \tag{2.8}$$

where $\hat{A} = \begin{bmatrix} 0 & I & 0 & \cdots & 0 \\ 0 & 0 & I & \cdots & 0 \\ \vdots & \vdots & \vdots & \ddots & \vdots \\ 0 & 0 & 0 & \cdots & I \\ \frac{A}{b_n} & -\frac{b_1}{b_n}I & -\frac{b_2}{b_n}I & \cdots & -\frac{b_{n-1}}{b_n}I \end{bmatrix}$, $\hat{B} = \begin{bmatrix} 0 \\ 0 \\ \vdots \\ 0 \\ \frac{B}{b_n} \end{bmatrix}$.

Now we have changed Case 4 to Case 3, which can be similarly analyzed.

The following can be obtained under the similar analysis procedure in the previous cases,

$$x(t) = L^{-1}\left[\left(\left(\int_0^1 \sum_{n=1}^N b_n \delta(\alpha - \beta_n) s^\alpha d\alpha\right) I - A\right)^{-1} B\right]$$

$$= L^{-1}\left[\left(\sum_{n=1}^N b_n s^{\beta_n} I - A\right)^{-1} B\right].$$

In the following, Case 4 will not be considered.

2.3 Time-Domain Analysis: Impulse Responses

Case 1 $w(\alpha) = 1$

As the transfer function of distributed-order system for Case 1 with the assumption that $D = 0$ is $G_1(s) = C \ln s((s-1)I - A \ln s)^{-1} B$, using the similar method of impulse response for distributed-order integrator/differentiator in Li et al. (2010), the inverse Laplace transform of $G_1(s)$ can be derived as follows.

$$y_1(t) = L^{-1}\{G_1(s)\}$$

$$= C\left(\frac{1}{2\pi i}\int_{\sigma-i\infty}^{\sigma+i\infty} e^{-st} \ln s(sI - (I + \ln sA))^{-1} ds\right) B$$

$$= C\left(\int_0^\infty e^{-xt}(x+1)A_1^{-1} dx\right) B \tag{2.9}$$

where $A_1 := ((x+1)I + A \ln x)^2 + (A\pi)^2$.

Case 2 $w(\alpha) = \alpha$

Following the same procedures, the transfer function of distributed-order system for Case 2 with $D = 0$ is $G_2(s) = C\ln^2 s((s-1-\ln s)I - A\ln^2 s)^{-1} B$, using the similar method of impulse response for distributed-order integrator/differentiator in Li et al. (2010), the inverse Laplace transform of $G_2(s)$ can be derived as follows.

$$y_2(t) = L^{-1}\{G_2(s)\}$$

$$= C\left(\frac{1}{2\pi i}\int_{\sigma-i\infty}^{\sigma+i\infty} e^{st}\ln^2 s\left((1-s+s\ln s)I - \ln^2 sA\right)^{-1} ds\right) B$$

$$= C\left(\int_0^\infty e^{-xt}\left(\begin{array}{c}((1+x-x\ln x)I + (\ln^2 x - \pi^2)A)^2 \\ +\pi^2(xI + 2\ln xA)^2\end{array}\right) A_2^{-1} dx\right) B$$

$$\tag{2.10}$$

where $A_2 := ((1+x-x\ln x)I + (\ln^2 x - \pi^2)A)^2 + \pi^2(xI + 2\ln xA)^2$.

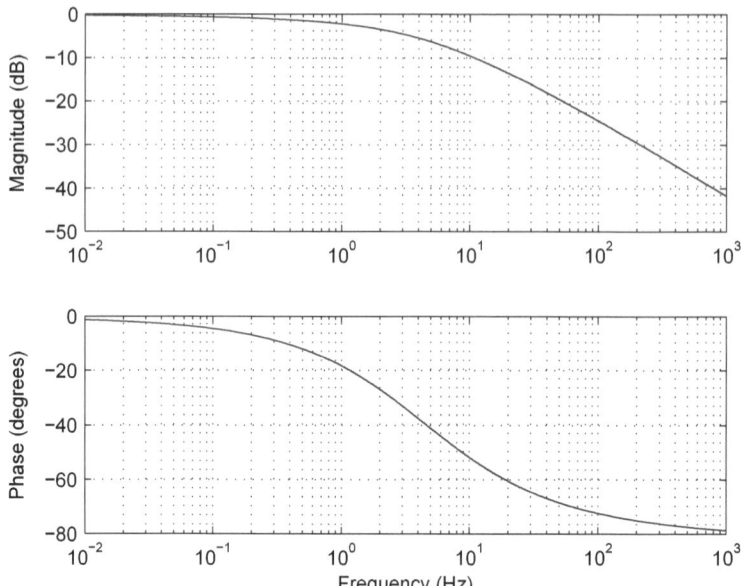

Fig. 2.6 The Bode plot of distributed-order system $g_2(s) = \frac{\ln^2 s}{1-s+s\ln s+\ln^2 s}$

Case 3 $w(\alpha) = \delta(\alpha - \beta), \ (0 < \beta < 1)$

The transfer function of fractional-order system for Case 3 with the assumption that $D = 0$ is $G_3(s) = C(s^\beta I - A)^{-1} B$, the inverse Laplace transform of $G_3(s)$ with null initial condition is

$$y_3(t) = C\left(t^{\beta-1} E_{\beta,\beta}(At^\beta)\right) B \qquad (2.11)$$

where $E_{\alpha,\beta}(\cdot)$ is the Mittag-Leffler function in two parameters defined as in Podlubny (1999)

$$E_{\alpha,\beta}(z) = \sum_{k=0}^{\infty} \frac{z^k}{\Gamma(\alpha k + \beta)}, \ (\Re(\alpha, \beta) > 0).$$

Remark 2.5 Computing (2.9), (2.10) and (2.11) can be easily realized in MATLAB numerically.

2.4 Frequency-Domain Response: Bode Plots

Generally, the frequency domain response has to be obtained by the direct evaluation of the irrational transfer function of distributed-order system along the imaginary axis for $s = j\omega$, $\omega \in (0, \infty)$. For simplicity, Bode plots of some scalar transfer functions for Case 1 to Case 3 are shown as follows.

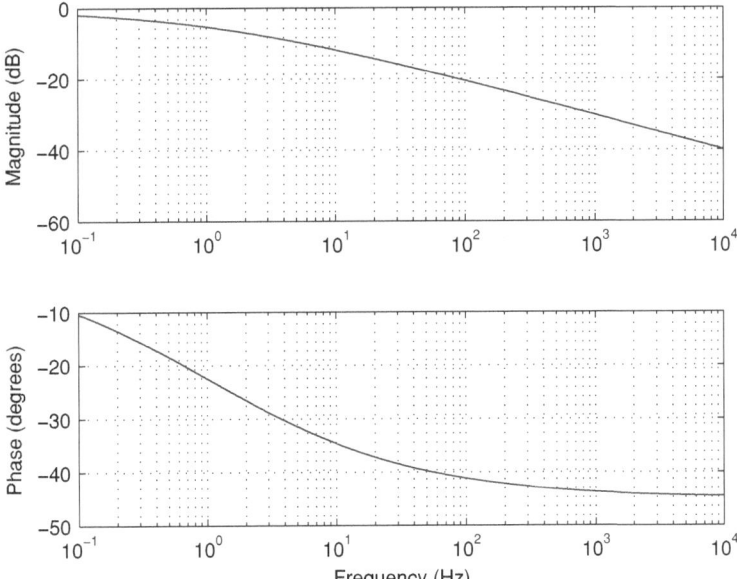

Fig. 2.7 The Bode plot of fractional-order system $g_3(s) = \frac{1}{s^{0.5}+1}$

For $w(\alpha) = 1$, the frequency-domain response of $g_1(s) = \frac{\ln s}{s-1+\ln s}$ is shown in Fig. 2.6.

For $w(\alpha) = \alpha$, the frequency-domain response of $g_2(s) = \frac{\ln^2 s}{1-s+s \ln s+\ln^2 s}$ is shown in Fig. 2.7.

For $w(\alpha) = \delta(\alpha - \beta)$, $(0 < \beta < 1)$, the frequency-domain response of $g_3(s) = \frac{1}{s^{0.5}+1}$ is shown in Fig. 2.8.

2.5 Numerical Examples

In this section, numerical examples are shown to demonstrate the effectiveness of the proposed results.

Example 1 Consider a distributed-order system with Case 1 described with parameters given as $A = \begin{bmatrix} 1 & 2 \\ -2 & 1 \end{bmatrix}$, $B = \begin{bmatrix} 1 \\ 1 \end{bmatrix}$, $C = \begin{bmatrix} 2 & 1 \end{bmatrix}$ and $D = 0$.

The eigenvalues of A are $\lambda_1 = 1 + 2j$ and $\lambda_2 = 1 - 2j$, so it can be known from Theorem 2.1 that this distributed-order system is bounded-input bounded-output stable. Using MATLAB to derive numerically, the states of impulse response with null initiations are shown in Figs. 2.9 and 2.10, respectively.

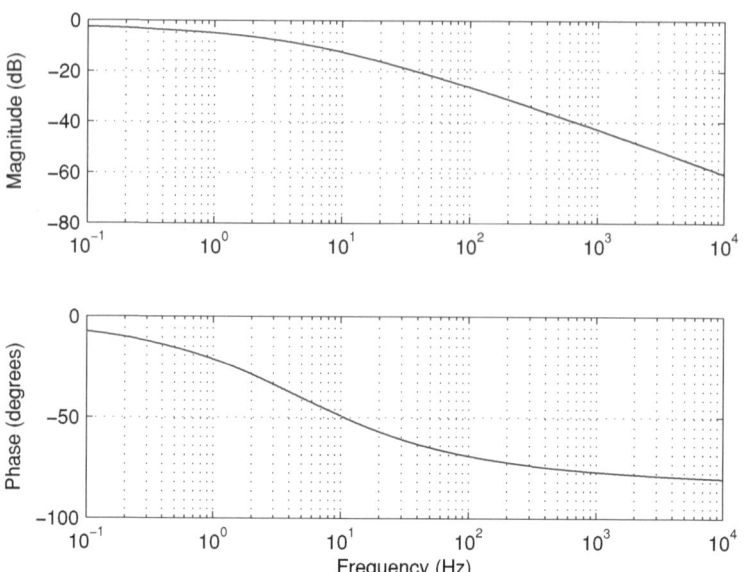

Fig. 2.8 The Bode plot of distributed-order system $g_1(s) = \frac{\ln s}{s - 1 + \ln s}$

Fig. 2.9 The state x_1 of stable distributed-order system (2.1) for Case 1

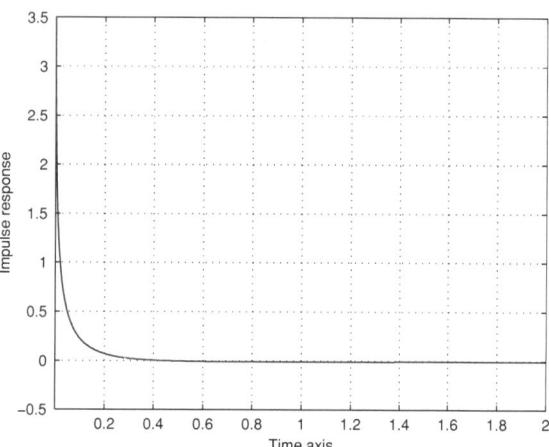

Example 2 Consider a distributed-order system with Case 1 described with parameters given as $A = \begin{bmatrix} 2 & 2 \\ -2 & 2 \end{bmatrix}$, $B = \begin{bmatrix} 1 \\ 1 \end{bmatrix}$, $C = \begin{bmatrix} 2 & 1 \end{bmatrix}$ and $D = 0$.

The eigenvalues of A are $\lambda_1 = 2 + 2j$ and $\lambda_2 = 2 - 2j$, and it can be known from Theorem 2.1 that this distributed-order system is not bounded-input bounded-output stable. Using MATLAB to derive numerically, the states of impulse response with null initiations are shown in Figs. 2.11 and 2.12, respectively.

Fig. 2.10 The state x_2 of stable distributed-order system (2.1) for Case 1

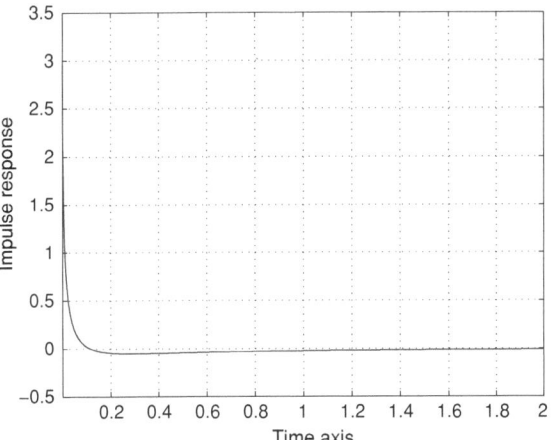

Fig. 2.11 The state x_1 of unstable distributed-order system (2.1) for Case 1

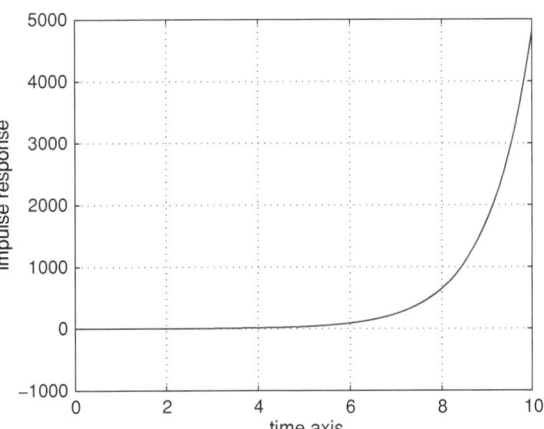

Example 3 Consider a distributed-order system with Case 2 described with parameters given as $A = \begin{bmatrix} 1 & 3 \\ -3 & 1 \end{bmatrix}$, $B = \begin{bmatrix} 1 \\ 1 \end{bmatrix}$, $C = \begin{bmatrix} 2 & 1 \end{bmatrix}$ and $D = 0$.

The eigenvalues of A are $\lambda_1 = 1 + 3j$ and $\lambda_2 = 1 - 3j$, so it can be known from Theorem 2.2 that this distributed-order system is bounded-input bounded-output stable, and the states of impulse response with null initiations are shown in Figs. 2.13 and 2.14, respectively.

Example 4 Consider a distributed-order system with Case 2 described with parameters given as $A = \begin{bmatrix} 2 & 2 \\ -2 & 2 \end{bmatrix}$, $B = \begin{bmatrix} 1 \\ 1 \end{bmatrix}$, $C = \begin{bmatrix} 2 & 1 \end{bmatrix}$ and $D = 0$.

The eigenvalues of A are $\lambda_1 = 2 + 2j$ and $\lambda_2 = 2 - 2j$, it can be known from Theorem 2.2 that this distributed-order system is not bounded-input bounded-output

Fig. 2.12 The state x_2 of unstable distributed-order system (2.1) for Case 1

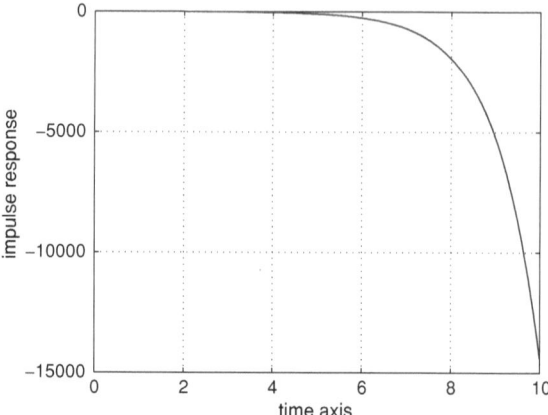

Fig. 2.13 The state x_1 of stable distributed-order system (2.1) for Case 2

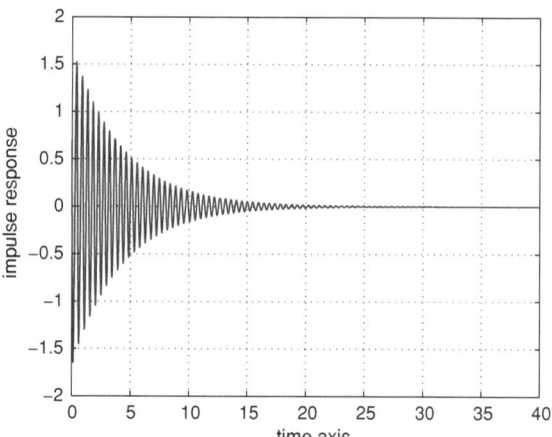

stable, and by using MATLAB to derive numerically, the states of impulse response with null initiations are shown in Figs. 2.15 and 2.16, respectively.

Example 5 Consider a fractional-order system for Case 3 described with parameters given as $\alpha = 0.5$, $A = \begin{bmatrix} 0 & 2 \\ -2 & 0 \end{bmatrix}$, $B = \begin{bmatrix} 1 \\ 1 \end{bmatrix}$, $C = \begin{bmatrix} 2 & 1 \end{bmatrix}$ and $D = 0$.

The eigenvalues of A are $\lambda_1 = 2j$ and $\lambda_2 = -2j$, it can be known from Theorem 2.3 that this fractional-order system is bounded-input bounded-output stable. Using MATLAB to derive numerically, the states of impulse response with null initiations are shown in Figs. 2.17 and 2.18, respectively.

Example 6 Consider a fractional-order system for Case 3 described with parameters given as $\alpha = 2/3$, $A = \begin{bmatrix} 1 & 1 \\ -1 & 1 \end{bmatrix}$, $B = \begin{bmatrix} 1 \\ 1 \end{bmatrix}$, $C = \begin{bmatrix} 2 & 1 \end{bmatrix}$ and $D = 0$.

Fig. 2.14 The state x_2 of stable distributed-order system (2.1) for Case 2

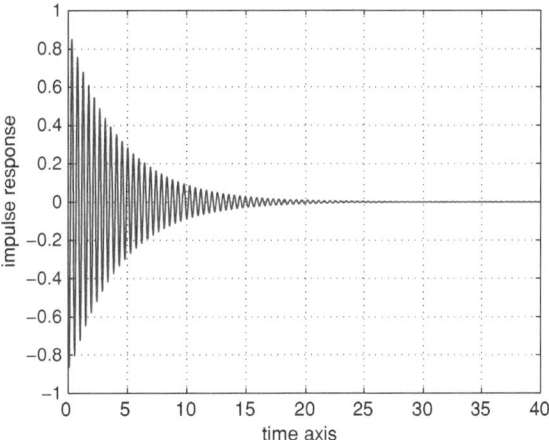

Fig. 2.15 The state x_1 of unstable distributed-order system (2.1) for Case 2

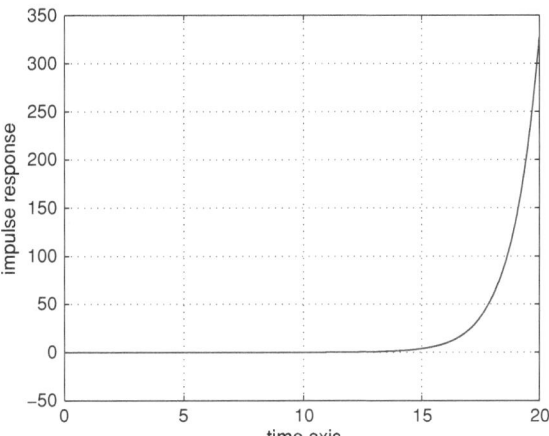

Since the eigenvalues of A are $\lambda_1 = 1 + j$ and $\lambda_2 = 1 - j$, it can be known from Theorem 2.3 that this fractional-order system is bounded-input bounded-output stable. Using MATLAB to derive numerically, the states of impulse response with null initiations are shown in Figs. 2.19 and 2.20, respectively.

2.6 Chapter Summary

In this chapter, the bounded-input bounded-output stability conditions for four kinds of linear time-invariant distributed-order system whose integral interval being $(0, 1)$ have been derived for the first time. Based on the final value property of Laplace transform, sufficient and necessary conditions of stability for distributed-order sys-

Fig. 2.16 The state x_2 of
unstable distributed-order
system (2.1) for Case 2

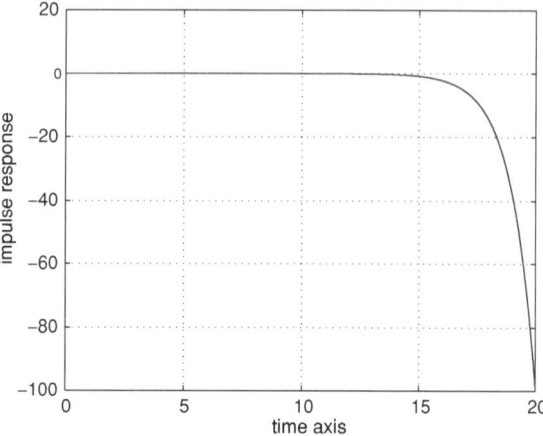

Fig. 2.17 The state x_1 of
stable fractional-order system
(2.5) for Case 3

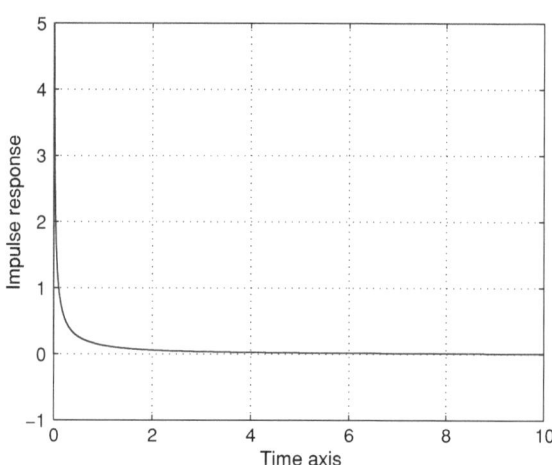

tems are presented. In addition, time-domain and frequency-domain responses are
presented with six illustrative numerical examples. Detailed MATLAB codes are
shown in Appendix A.

Fig. 2.18 The state x_2 of stable fractional-order system (2.5) for Case 3

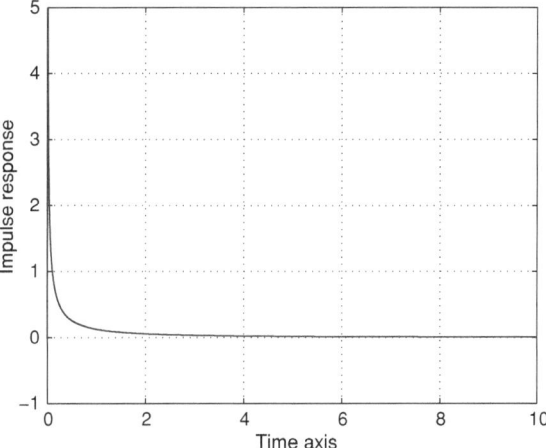

Fig. 2.19 The state x_1 of unstable fractional-order system (2.5) for Case 3

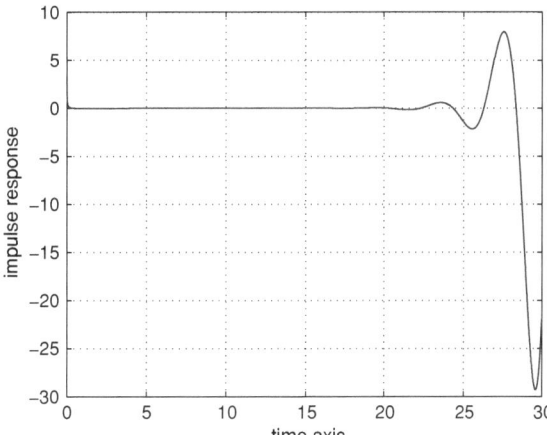

Fig. 2.20 The state x_2 of
unstable fractional-order
system (2.5) for Case 3

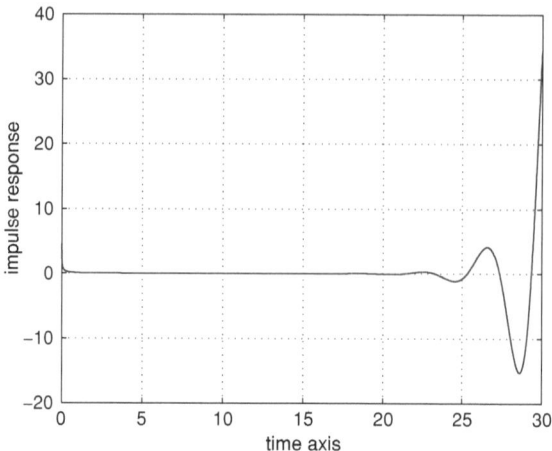

References

Asmar NH, Jones GC (2002) Applied complex analysis with partial differential equations. Prentice
    Hall, Upper Saddle River
Cuadrado M, Cabanes R (1989) T. de Variable Compleja. Servicio de Publicaciones de la ETSIT
    UPM, Madrid
Gross B, Braga EP (1961) Singularities of linear system functions. Elsevier, New York
Hartley TT, Lorenzo CF (1999) Fractional system identification: an approach using continuous
    order-distributions. NASA Tech Memo 209640:20
Hartley TT, Lorenzo CF (2003) Fractional-order system identification based on continuous order-
    distributions. Signal Process 83(11):2287–2300
Hartley TT, Lorenzo CF (2004) A frequency-domain approach to optimal fractional-order damping.
    Nonlinear Dyn 38(1–2):69–84
Li Y, Sheng H, Chen YQ (2010) On distributed order integrator/differentiator. Signal Process
    91(5):1079–1084
Matignon D (1996) Stability results on fractional differential equations with applications to con-
    trol processing. In: Multiconference on computational engineering in systems and application,
    pp 963–968
Podlubny I (1999) Fractional differential equations. Academic Press, San Diego
Westerlund S, Ekstam L (1994) Capacitor theory. IEEE Trans Dielectr Electr Insul 1(5):826–839

Chapter 3
Noncommensurate Constant Orders as Special Cases of DOLTIS

3.1 Introduction

Stability is a minimum requirement for control systems, certainly including fractional-order systems. In Matignon (1996), the stability results on fractional-order linear time-invariant (FO-LTI) systems with commensurate orders were presented for the first time, it permits to check the asymptotically stability through the location of the system matrix eigenvalues of the pseudo state space representation of fractional-order system in the Complex plane. Henceforth, there were some systematic results on the robust stability of interval uncertain FO-LTI systems as presented in Ahn and Chen (2008), Ahn et al. (2007), Chen et al. (2006), Lu and Chen (2010), Petras et al. (2004). The BIBO-stability of fractional-order delay systems of retarded and neutral types was studied in Bonnet and Partington (2002), in which necessary and sufficient conditions were presented for retarded type, and only sufficient conditions were provided for neutral type. In Bonnet and Partington (2007), necessary and sufficient conditions of stability were provided for an important special case fractional-order delay system of neutral type. However, such theorems obtained in Bonnet and Partington (2002, 2007) don't permit to conclude the system stability without computing the system's poles, which constitutes tedious work, so based on Cauchy's integral theorem and by solving an initial-value problem, an effective numerical algorithm for testing the BIBO stability of fractional delay systems was presented in Hwang and Cheng (2006).

However, the fractional-order systems discussed in these literatures are mostly with commensurate orders, which means the orders can always converted to commensurate orders when they have a common divisor. To the best of our knowledge, there are few results concerning the stability analysis problems for fractional-order systems with noncommensurate orders. Based on Cauchy's theorem, a graphical test to evaluate fractional-order systems with noncommensurate orders are given in Sabatier et al. (2010), however, this method is not very helpful because of the complicated procedures. Therefore motivated by the previous references, this section

Z. Jiao et al., *Distributed-Order Dynamic Systems,* SpringerBriefs in Control, Automation and Robotics, DOI: 10.1007/978-1-4471-2852-6_3,

addresses the bounded-input bounded-output stability for fractional-order systems with multiple discrete noncommensurate orders.

3.2 Stability Analysis of Some Special Cases of DOLTIS

3.2.1 Case 1: Double Noncommensurate Orders

For the distributed-order system with double noncommensurate orders described by

$$_0D_t^{w(\alpha)}x(t) = \int_0^1 w(\alpha)_0D_t^\alpha x(t)d\alpha = Ax(t) + Bu(t)$$

$$y(t) = Cx(t) + Du(t), \tag{3.1}$$

where $_0D_t^\alpha$ denotes Caputo fractional-order derivative operator, $w(\alpha) = \delta(\alpha - \beta_1) + \delta(\alpha - \beta_2)$ is the function distribution of order $\alpha \in [0, 1]$, $0 < \beta_1$, $\beta_2 \le 1$ are noncommensurate orders, which means that they do not have a common divisor.

Remark 3.1 When the function distribution of order takes discrete values, distributed-order system (3.1) will be fractional-order system with double noncommensurate orders. Fractional-order system (3.1) can always be converted to a fractional-order system with commensurate orders if both β_1 and β_2 are rational numbers (Monje et al. 2010), and the stability issue of fractional-order system with commensurate orders has been solved in Matignon (1996). When at least one of β_1 and β_2 is not a rational number, or β_1 and β_2 are not commensurate numbers, they do not have common divisors. Based on Cauchy's theorem, a graphical test to evaluate fractional-order systems with noncommensurate orders are given in Sabatier et al. (2010), and the system is considered in frequency domain, however, this method is not very helpful because of the complicated procedures. In current section, sufficient and necessary condition for fractional-order system with double noncommensurate orders is proposed first.

Under the assumption of zero initial conditions, taking the Laplace transform of (3.1), we have

$$s^{\beta_1}X(s) + s^{\beta_2}X(s) = AX(s) + BU(s).$$

Assume that $D = 0$. The transfer function of (3.1) is

$$H(s) = C\left((s^{\beta_1} + s^{\beta_2})I - A\right)^{-1}B.$$

Similar to the BIBO stability for traditional control systems, we have the following definition.

Fig. 3.1 The stable boundary of fractional-order system (3.1) with double noncommensurate orders $\beta_1 = \sqrt{2}-1$ and $\beta_2 = \sqrt{3}-1$

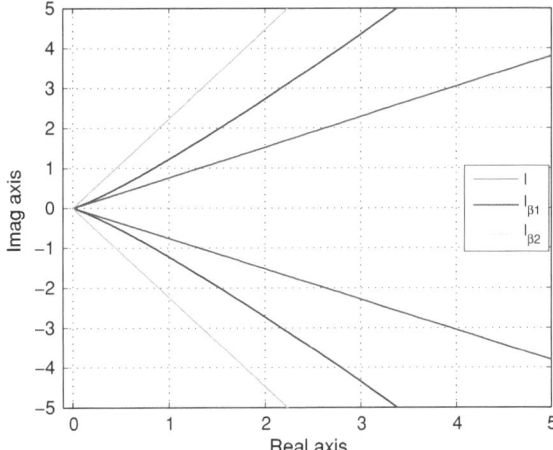

Definition 3.1 Fractional-order system (3.1) defined by its impulse response $h(t) = L^{-1}\{H(s)\}$ is BIBO stable if and only if $\forall u \in L^{\infty}(R^+)$, $h * u \in L^{\infty}(R^+)$, where $*$ stands for the convolution product and $L^{\infty}(R^+)$ stands for the Lebesgue space of measurable function h such that $ess \sup_{t \in R^+} |h(t)| < \infty$.

Theorem 3.1 *Fractional-order system (3.1) is BIBO stable if and only if all the eigenvalues of A lie on the left of curve* $l_4 := l_g \bigcup l_h$, *where* l_g *and* l_h *are symmetrical with respect to the real axis, and* $l_g := \{x + iy \,|\, x = x_\omega, y = y_\omega, \omega \in [0, \infty)\}$, *with* x_ω *and* y_ω *defined as* $x_\omega := \omega^{\beta_1} \cos \frac{\beta_1}{2}\pi + \omega^{\beta_2} \cos \frac{\beta_2}{2}\pi$, $y_\omega := \omega^{\beta_1} \sin \frac{\beta_1}{2}\pi + \omega^{\beta_2} \sin \frac{\beta_2}{2}\pi$.

Proof The proof of this theorem can be followed by the similar proof procedures of Theorem 2.1. The stable boundary of fractional-order system (3.1) l_4 is plotted in Fig. 3.1, with the local property around 0 zoomed in Fig. 3.2 with $\beta_1 = \sqrt{2} - 1$ and $\beta_2 = \sqrt{3} - 1$.

Generally, the frequency domain response can be obtained by the direct evaluation of the irrational transfer function of fractional-order system with double noncommensurate orders along the imaginary axis for $s = j\omega$, $\omega \in (0, \infty)$. In the following, the Bode plot of fractional-order system (3.1) with $\beta_1 = \sqrt{2} - 1$ and $\beta_2 = \sqrt{3} - 1$ is shown in Fig. 3.3.

Remark 3.2 For fractional-order system described by $_0D_t^{\beta_i} x(t) = Ax(t) + Bu(t)$ $(i = 1, 2)$, the stable boundary is defined as $l_{\beta_i} := l_{a_i} \bigcup l_{b_i}$, where l_{a_i} and l_{b_i} are symmetrical with respect to the real axis, and l_{a_i} is defined as:

$$l_{a_i} := \left\{ x + iy \,\bigg|\, x = \omega^{\beta_i} \cos \frac{\beta_i}{2}\pi, y = \omega^{\beta_i} \sin \frac{\beta_i}{2}\pi, \omega \in [0, \infty). \right\}$$

Fig. 3.2 The stable boundary of fractional-order system (3.1) with double noncommensurate orders $\beta_1 = \sqrt{2}-1$ and $\beta_2 = \sqrt{3}-1$ (*Zoomed*)

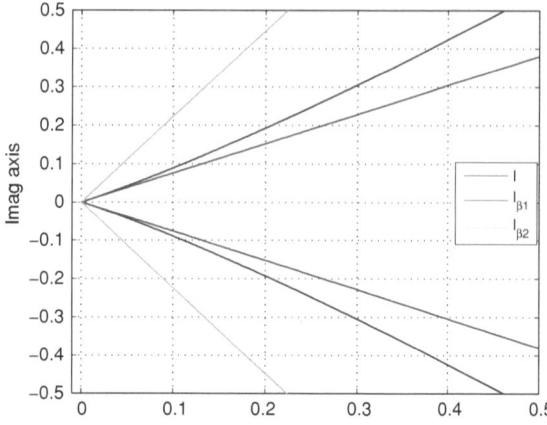

Fig. 3.3 The bode plot of fractional-order system (3.1) with double noncommensurate orders $\beta_1 = \sqrt{2}-1$ and $\beta_2 = \sqrt{3}-1$

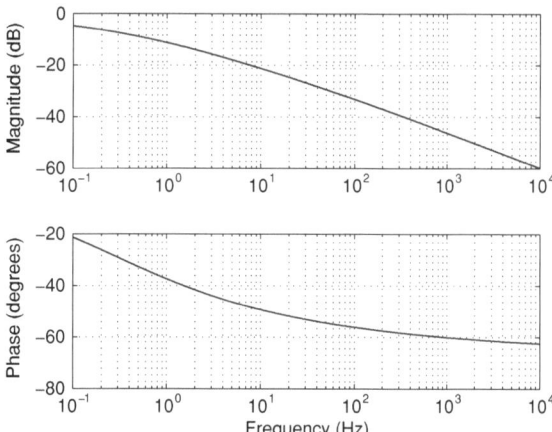

l_{β_1} and l_{β_2} are also plotted in Figs. 3.1 and 3.2. It can be easily seen from Figs. 3.1 and 3.2 that curve l lies between l_{β_1} and l_{β_2}.

Remark 3.3 We assume that $0 < \beta_1, \beta_2 \leq 1$ in (3.1), which means that both β_1 and β_2 cannot be 0. Without loss of generality, if $\beta_2 = 0$, β_1 is an irrational number, fractional-order system (3.1) will become $_0D_t^{\beta_1}x(t) = \hat{A}x(t) + Bu(t)$, with $\hat{A} = A - I$. In this case, the stable boundary is l_{β_1} defined in Remark 3.2 with respect to $\hat{A}$.

Remark 3.4 If both β_1 and β_2 are irrational numbers. Let $\beta_2 = k\beta_1$, where k is a positive integer. Here it means that the orders are commensurate orders, then the following cases can be easily given:

- If $k = 1$, fractional-order system (3.1) will become $2_0 D_t^{\beta_1} x(t) = Ax(t) + Bu(t)$, i.e., $_0 D_t^{\beta_1} x(t) = A_1 x(t) + B_1 u(t)$, with $A_1 = A/2$, $B_1 = B/2$, and the stable boundary is l_{β_1} defined in Remark 3.2 with respect to A_1.
- If $k = 2$, fractional-order system (3.1) will become $_0 D_t^{2\beta_1} x(t) + _0 D_t^{\beta_1} x(t) = Ax(t) + Bu(t)$. Let $\hat{x}(t) := \left[x(t)\ _0 D_t^{\beta_1} x(t) \right]^T$, then we have $D^{\beta_1} \hat{x}(t) = A_2 \hat{x}(t) + B_2 u(t)$, with $A_2 = \begin{bmatrix} 0 & I \\ A & -I \end{bmatrix}$, $B_2 = \begin{bmatrix} 0 \\ B \end{bmatrix}$, and the stable boundary is l_{β_1} defined in Remark 3.2 with respect to A_2.
- If k is any positive integer, the similar conclusion can be obtained.

3.2.2 Case 2: N-Term Noncommensurate Orders

For the fractional-order system with N-term noncommensurate orders described by

$$_0 D_t^{w(\alpha)} x(t) = \int_0^1 w(\alpha)_0 D_t^{\alpha} x(t) d\alpha = Ax(t) + Bu(t)$$

$$y(t) = Cx(t) + Du(t) \qquad (3.2)$$

where $_0 D_t^{\alpha}$ denotes Caputo fractional-order derivative operator, $w(\alpha) = \sum_{i=1}^n \delta(\alpha - \beta_i)$ is the function distribution of order $\alpha \in [0, 1]$, $0 < \beta_1, \beta_2, \cdots, \beta_n \leq 1$ are N-term noncommensurate orders, which means that they do not have a common divisor.

Similar to the analysis in Case 1, let $u(t) = \delta(t)$, then the transfer function of (3.2) under the assumption of zero initial conditions is

$$H_1(s) = C \left(\sum_{i=1}^n s^{\beta_i} I - A \right)^{-1} B.$$

We have the following parallel result.

Theorem 3.2 *Fractional-order system (3.2) is BIBO stable if and only if all the eigenvalues of A lie on the left of curve $l_5 := l_i \bigcup l_j$, where l_i and l_j are symmetrical with respect to the real axis, and l_i is defined as:*

$$l_i := \{x + iy \,|\, x = x_\omega, y = y_\omega, \omega \in (0, \infty)\}$$

with x_ω and y_ω defined as $x_\omega := \sum_{i=1}^n \omega^{\beta_i} \cos \frac{\beta_i}{2}\pi$, $y_\omega := \sum_{i=1}^n \omega^{\beta_i} \sin \frac{\beta_i}{2}\pi$.

Proof The proof of this theorem is similar to the proof of Theorem 2.1.

Fig. 3.4 The state x_1 of stable fractional-order system (3.1) with double noncommensurate orders $\beta_1 = \sqrt{2} - 1$ and $\beta_2 = \sqrt{3} - 1$

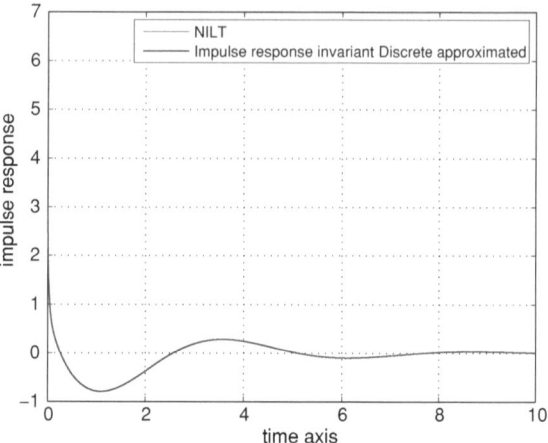

Remark 3.5 From the calculation and plots through MATLAB in Case 1, it can be known that curve l_5 must lies between l_{min} and l_{max} which are defined as

$$l_{min} := \left\{ re^{i\theta} \,\middle|\, \theta = \pm \frac{\min\{\beta_i\}}{2}\pi, r \geq 0 \right\}$$

and

$$l_{max} := \left\{ re^{i\theta} \,\middle|\, \theta = \pm \frac{\max\{\beta_i\}}{2}\pi, r \geq 0 \right\}.$$

3.3 Numerical Examples

In this section, numerical examples are presented to demonstrate the effectiveness of the proposed concept.

Example 1 Consider a fractional-order system (3.1) with double noncommensurate orders described with parameters $\beta_1 = \sqrt{2} - 1$, $\beta_2 = \sqrt{3} - 1$, $A = \begin{bmatrix} 1 & 2 \\ -2 & 1 \end{bmatrix}$, $B = \begin{bmatrix} 1 \\ 1 \end{bmatrix}$, $C = \begin{bmatrix} 2 & 1 \end{bmatrix}$ and $D = 0$.

The eigenvalues of A are $\lambda_1 = 1 + 2j$ and $\lambda_2 = 1 - 2j$, and it can be known from Theorem 3.1 that this fractional-order system is bounded-input bounded-output stable. Based on the numerical inverse Laplace transform (NILT) technique (Li et al. 2011), the states of impulse response for $G_{d1}(s) = C\left((s^{\beta_1} + s^{\beta_2})I - A\right)^{-1}B$ with null initiations are shown in Figs. 3.4 and 3.5, respectively.

Fig. 3.5 The state x_2 of stable fractional-order system (3.1) with double noncommensurate orders $\beta_1 = \sqrt{2} - 1$ and $\beta_2 = \sqrt{3} - 1$

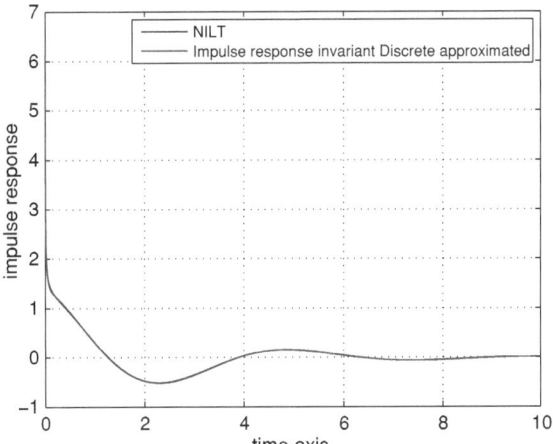

Fig. 3.6 The state x_1 of unstable fractional-order system (3.1) with double noncommensurate orders $\beta_1 = \sqrt{2}-1$ and $\beta_2 = \sqrt{3}-1$

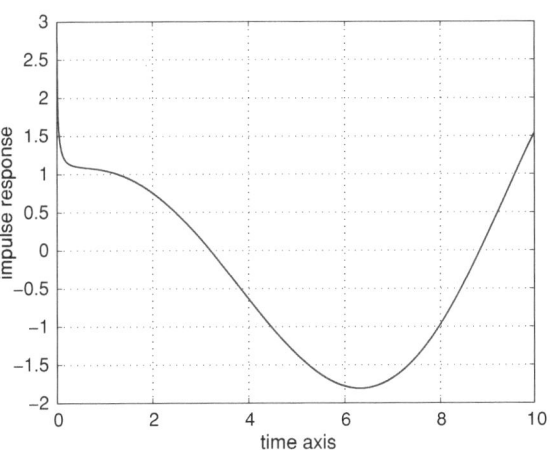

Example 2 Consider a fractional-order system (3.1) with double noncommensurate orders described with parameters $\beta_1 = \sqrt{2} - 1$, $\beta_2 = \sqrt{3} - 1$, $A = \begin{bmatrix} 1 & 1 \\ -1 & 1 \end{bmatrix}$, $B = \begin{bmatrix} 1 \\ 1 \end{bmatrix}$, $C = \begin{bmatrix} 2 & 1 \end{bmatrix}$ and $D = 0$.

The eigenvalues of A are $\lambda_1 = 1 + j$ and $\lambda_2 = 1 - j$, so it can be known from Theorem 3.1 that this fractional-order system is not bounded-input bounded-output stable. Based on the NILT technique, the states of impulse response for $G_{d2} = C\big((s^{\beta_1} + s^{\beta_2})I - A\big)^{-1}B$ with null initiations are shown in Figs. 3.6 and 3.7, respectively.

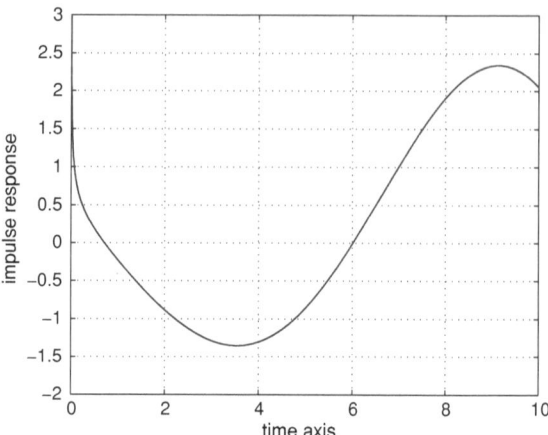

Fig. 3.7 The state x_2 of unstable fractional-order system (3.1) with double noncommensurate orders $\beta_1 = \sqrt{2} - 1$ and $\beta_2 = \sqrt{3} - 1$

3.4 Chapter Summary

In this chapter, the stability issues for fractional-order linear time-invariant systems with multiple noncommensurate orders are solved. The double noncommensurate orders and N-term noncommensurate orders are analyzed respectively and sufficient and necessary conditions of stability are obtained. In addition, based on the numerical inverse Laplace transform technique, time-domain responses for the double noncommensurate order case are presented to verify the main results as illustrative numerical examples. Detailed MATLAB codes are presented in Appendix A.

References

Ahn HS, Chen YQ (2008) Necessary and sufficient stability condition of fractional-order interval linear systems. Automatica 44(11):2985–2988

Ahn HS, Chen YQ, Podlubny I (2007) Robust stability test of a class of linear time-invariant interval fractional-order systems using Lyapunov inequality. Appl Math Comput 187(1):27–34

Bonnet C, Partington JR (2002) Analysis of fractional delay systems of retarded and neutral type. Automatica 38(7):1133–1138

Bonnet C, Partington JR (2007) Stabilization of some fractional delay systems of neutral type. Automatica 43(12):2047–2053

Chen YQ, Ahn HS, Podlubny I (2006) Robust stability check of fractional order linear time invariant systems with interval uncertainties. Signal Process 86(10):2611–2618

Hwang C, Cheng YC (2006) A numerical algorithm for stability testing of fractional delay systems. Automatica 42(5):825–831

Li Y, Sheng H, Chen YQ (2011) Analytical impulse response of a fractional second order filter and its impulse response invariant discretization. Signal Process 91(3):498–507

Lu JG, Chen YQ (2010) Robust stability and stabilization of fractional order interval systems with the fractional order α: the $0 < \alpha < 1$ case. IEEE Trans Autom Control 55(1):152–158

Matignon D (1996) Stability results on fractional differential equations with applications to control processing. In: Multiconference on computational engineering in systems and application, pp 963–968

Monje CA, Chen YQ, Vinagre BM, Xue DY, Feliu V (2010) Fractional-order systems and controls: fundamentals and applications. Springer, London

Petras I, Chen YQ, Vinagre BM, Podlubny I (2004) Stability of linear time invariant systems with interval fractional orders and interval coefficients. In: Proceedings of the international conference on computational cybernetics, pp 341–346

Sabatier J, Farges C, Trigeassou JC (2010) A stability test for non commensurate fractional order system. In: Proceedings of the 4th IFAC workshop on fractional differentiation and its applications

Chapter 4
Distributed-Order Filtering and Distributed-Order Optimal Damping

The idea of using the distributed-order differential equation first proposed by Caputo (1969) is at least mathematically interesting as demonstrated in the previous chapters. However, people may question its usefulness in engineering practice. In this chapter, we included two generic applications. One is on distributed order signal processing and the other is on optimal distributed damping. We hope these two initial applications can serve as motivating examples to further the investigation in distributed order dynamics systems, signal processing, modeling and controls.

4.1 Application I: Distributed-Order Filtering

Filtering as a special type of dynamic system is a signal processing technique. In this section, we present two distributed-order filters: the distributed-order integrator/differentiator and the distributed-order low-pass filter. These two distributed order filters are studied in both time domain and frequency domain. Moreover, the discretization method is used to obtain the digital impulse responses of these distributed-order fractional filters. The results are verified in both time and frequency domains.

4.1.1 Distributed-Order Integrator/Differentiator

Motivated by the applications of the distributed-order operators in control, filtering and signal processing, a distributed-order integrator/differentiator is derived step by step in this section. Firstly, the classical integer order integrator can be rewritten as

$$\frac{1}{s} = \int_{-\infty}^{\infty} \delta(\alpha - 1) \frac{1}{s^{\alpha}} d\alpha, \tag{4.1}$$

Z. Jiao et al., *Distributed-Order Dynamic Systems,* SpringerBriefs in Control, Automation and Robotics, DOI: 10.1007/978-1-4471-2852-6_4,
© The Author(s) 2012

where $\delta(\cdot)$ denotes the Dirac–Delta function and $\frac{1}{s^\alpha}$ is the fractional-order integrator/differentiator with order $\alpha \in \mathbb{R}$. Moreover, the summation of a series of fractional-order integrators/differentiators can be expressed as

$$\sum_k \frac{1}{s^{\alpha_k}} = \int_{-\infty}^{\infty} \left[\sum_k \delta(\alpha - \alpha_k)\right] \frac{1}{s^\alpha} d\alpha, \tag{4.2}$$

where k can belong to any countable or noncountable set. Now, it is straightforward to replace $\left[\sum_k \delta(\alpha - \alpha_k)\right]$ by a weighted kernel function $w(\alpha)$. It follows that the right side of the above equation becomes

$$\int_{-\infty}^{\infty} w(\alpha) \frac{1}{s^\alpha} d\alpha, \tag{4.3}$$

where $w(\alpha)$ is independent of time, and the above equation defines a distributed-order integrator/differentiator. Particularly, when $w(\alpha)$ is a piecewise function,

$$\int_{-\infty}^{\infty} w(\alpha) \frac{1}{s^\alpha} d\alpha = w(\alpha_l) \int_{a_l}^{b_l} \frac{1}{s^\alpha} d\alpha, \tag{4.4}$$

where a_l, b_l are real numbers, $\alpha_l \in (a_l, b_l)$ and $w(\alpha)$ is a constant on $\alpha \in (a_l, b_l)$. Based on the above discussions, without loss of generality, we focus on the uniform distributed-order integrator/differentiator $\int_a^b \frac{1}{s^\alpha} d\alpha$, where $a < b$ are arbitrary real numbers.

In order to apply the distributed-order integrator/differentiator, the numerical discretization method is needed. This finds applications in signal modeling, filter design, controller design (Machado 1997) and nonlinear system identification (Hartley and Lorenzo 2003; Adams et al. 2008). The numerical discretization of the distributed-order integrator/differentiator, the key step towards application, can be realized in two ways: direct methods and indirect methods. In indirect discretization methods (Oustaloup et al. 2000; Chen and Moore 2002), two steps are required, i.e., frequency domain fitting in continuous time domain first and then discretizing the fit s-transfer function (Chen and Moore 2002). Other frequency-domain fitting methods can also be used but without guaranteeing the stable minimum-phase discretization (Chen and Moore 2002). In this section, the direct discretization method will be used by an effective impulse response invariant discretization method discussed in Chen and Moore (2002), Chen and Vinagre (2003), Lubich (1986), Chen et al. (2004), and Li et al. (2011). In the above-mentioned references, the authors developed a technique for designing the discrete-time IIR filters from continuous-time fractional-order filters, in which the impulse response of the continuous-time fractional-order filter is sampled to produce the impulse response of the discrete-time filter. The detailed techniques of the impulse response invariant discretization method will be introduced in Sect. 4.1.1.2. For more discussions of the discretization methods, we cite Vinagre et al. (2003), Barbosa and Machado (2006), Ferdi (2009), Oustaloup (1981) and Radwan et al. (2007, 2008).

4.1.1.1 Impulse Response of the Distributed-Order Integrator/Differentiator

For the distributed-order integrator/differentiator (DOI/D)

$$\int_a^b \frac{1}{s^\alpha} d\alpha = \frac{s^{-a} - s^{-b}}{\ln(s)}, \tag{4.5}$$

where $a < b$ are arbitrary real numbers, its inverse Laplace transform is written as

$$\mathcal{L}^{-1}\left\{\int_a^b \frac{1}{s^\alpha} d\alpha\right\} = \frac{1}{2\pi i} \int_{\sigma-i\infty}^{\sigma+i\infty} e^{st} \frac{s^{-a} - s^{-b}}{\ln(s)} ds, \tag{4.6}$$

where $\sigma > 0$. It can be seen that there are two branch points of (4.5), $s = 0$ and $s = \infty$. Therefore, we can cut the complex plane by connecting the branch points along the negative real domain, so that the path integral in (4.6) is equivalent to the path integral along the Hankel path.[1] The Hankel path starts from $-\infty$ along lower side of real (horizontal) axis, encircles the circular disc $|s| = \varepsilon \to 0$, in the positive sense, and ends at $-\infty$ along the upper side of real axis. Moreover, it can also be proved that the path integral of $\frac{e^{st}(s^{-a}-s^{-b})}{\ln(s)}$ along $s \to 0$ equals zero for $b \leq 1$, and that there are no poles in the single value analytical plane. Therefore, by substituting $s = -xe^{-i\pi}$ and $s = xe^{i\pi}$, where $x \in (0, +\infty)$, we have, for an arbitrary $\sigma > 0$ and $b \leq 1$,

$$\mathcal{L}^{-1}\left\{\int_a^b \frac{1}{s^\alpha} d\alpha\right\} = \frac{1}{2\pi i} \int_{\sigma-i\infty}^{\sigma+i\infty} \frac{e^{st}\left(s^{-a} - s^{-b}\right)}{\ln(s)} ds$$

$$= \frac{1}{2\pi i} \int_0^\infty \frac{e^{-xt}\left(x^{-a}e^{a\pi i} - x^{-b}e^{b\pi i}\right)}{\ln(x) - i\pi} dx$$

$$- \frac{1}{2\pi i} \int_0^\infty \frac{e^{-xt}\left(x^{-a}e^{-a\pi i} - x^{-b}e^{-b\pi i}\right)}{\ln(x) + i\pi} dx$$

$$= \frac{1}{\pi} \int_0^\infty \frac{e^{-xt}}{(\ln(x))^2 + \pi^2}\Big[x^{-a}(\sin(a\pi)\ln(x) + \pi\cos(a\pi))$$

$$- x^{-b}(\sin(b\pi)\ln(x) + \pi\cos(b\pi))\Big] dx. \tag{4.7}$$

Based on the above discussions, we arrive at the following theorem.

---

[1] It follows from the residue of $\frac{e^{st}(s^{-a}-s^{-b})}{\ln(s)}$ which equals zero at $s = \infty$, that the path integral of it along $s \to \infty$ is vanished for $b \leq 1$.

Theorem 4.1 *For any $a < b \leq 1$, we have*

$$\mathcal{L}^{-1}\left\{\int_a^b \frac{1}{s^\alpha}\mathrm{d}\alpha\right\} = \frac{1}{\pi}\int_0^\infty \frac{e^{-xt}}{(\ln(x))^2 + \pi^2}\Big[x^{-a}(\sin(a\pi)\ln(x) + \pi\cos(a\pi))$$
$$- x^{-b}(\sin(b\pi)\ln(x) + \pi\cos(b\pi))\Big]\mathrm{d}x. \qquad (4.8)$$

Especially, when $0 \leq a < b \leq 1$, it can be derived that

$$\left|\mathcal{L}^{-1}\left\{\int_a^b \frac{1}{s^\alpha}\mathrm{d}\alpha\right\}\right| \leq \frac{1}{\pi^2}\left(\frac{M_1 t^{a-1}}{|a-1|} + \frac{M_2 t^{b-1}}{|b-1|}\right), \qquad (4.9)$$

where M_1 and M_2 are finite positive constants.

Proof The first equation in this theorem is the same as (4.7). Moreover, by using (4.7), it can be easily proved that

$$\left|\mathcal{L}^{-1}\left\{\int_a^b \frac{1}{s^\alpha}\mathrm{d}\alpha\right\}\right| \leq \frac{1}{\pi^2}\int_0^\infty e^{-xt}(x^{-a} + x^{-b})\mathrm{d}x = \frac{1}{\pi^2}\left(\frac{M_1 t^{a-1}}{|a-1|} + \frac{M_2 t^{b-1}}{|b-1|}\right),$$

where $M_1 = \int_0^\infty e^{-\tau^{1/(1-a)}}\mathrm{d}\tau$ and $M_2 = \int_0^\infty e^{-\tau^{1/(1-b)}}\mathrm{d}\tau$ are finite positive constants for any $0 \leq a < b \leq 1$.

Based on the above discussions we can get the time domain expression of the impulse response of the distributed-order integrator/differentiator for any $a < b \leq 1$. Note here, for $a < b \leq 1$, (4.7) can be easily computed by using *"quadgk"* in MATLAB®, which will be used in the discretization method. Moreover, in order to extend a and b to the whole real axis, we can use the following properties.

Property 4.1 *It can be proved that*

(A) $s^c \int_a^b \frac{1}{s^\alpha}\mathrm{d}\alpha = \int_{a-c}^{b-c} \frac{1}{s^\alpha}\mathrm{d}\alpha$, *where $c \in \mathbb{R}$.*

(B) $\int_a^b \frac{1}{s^\alpha}\mathrm{d}\alpha = s\int_{a+1}^1 \frac{1}{s^\alpha}\mathrm{d}\alpha + \int_0^b \frac{1}{s^\alpha}\mathrm{d}\alpha$, *where $a \in [-1, 0)$ and $b \in [0, 1]$.*

(C) $\int_a^b \frac{1}{s^\alpha}\mathrm{d}\alpha = (s^{-1} + \cdots + s^{-N})s^{-[a]}\int_{a-[a]-1}^{a-[a]} \frac{1}{s^\alpha}\mathrm{d}\alpha + s^{-(N+[a]+1)} \cdot \int_{a-[a]-1}^{b-(N+[a]+1)} \frac{1}{s^\alpha}\mathrm{d}\alpha$, *where $b - a > 1$, $N = [b - a]$ and $[*]$ denotes the integer part of $*$.*

(D) $\int_{\tilde{a}}^{\tilde{b}} s^\alpha \mathrm{d}\alpha = s\int_a^b \frac{1}{s^\alpha}\mathrm{d}\alpha$, *where $\tilde{a} < \tilde{b}$, $a = 1 - \tilde{b}$ and $b = 1 - \tilde{a}$.*

(E) *The distributed integrator/differentiator $\int_a^b \frac{w(\alpha)}{s^\alpha}\mathrm{d}\alpha$, where $w(\alpha)$ is a piecewise function, can be converted to the summation of uniformly distributed integrators/differentiators.*

Theorem 4.2 *Any distributed-order integrator/differentiator can be composed by the distributed-order integrator for $0 \leq a < b \leq 1$, integrator $\frac{1}{s}$ and differentiator s.*

Proof This theorem can be proved by Property 4.1.

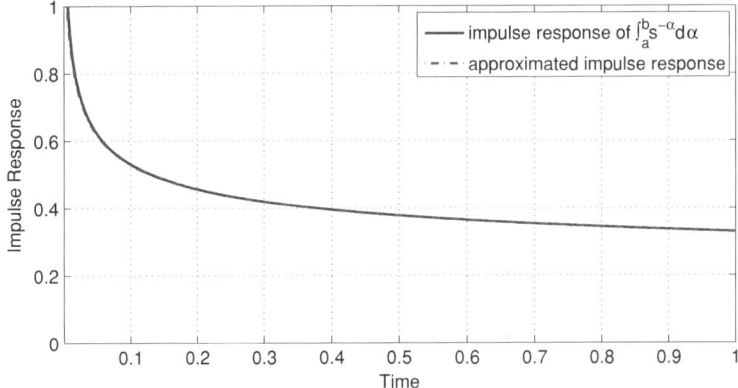

Fig. 4.1 The impulse responses of the approximate discrete-time IIR filter and the continuous-time distributed-order filter when $a = 0.6$, $b = 1$ and $T_s = 0.001\,\mathrm{s}$

4.1.1.2 Impulse Response Invariant Discretization of DOI/DOD

The impulse response invariant discretization method converts analog filter transfer functions to digital filter transfer function in such a way that the impulse responses are the same (invariant) at the sampling instants. Thus, if $g(t)$ denotes the impulse-response of an analog (continuous-time) filter, then the digital (discrete-time) filter given by the impulse-invariant method will have impulse response $g(nT_s)$, where T_s denotes the sampling period in seconds.

Impulse response invariance-based IIR-type discretization method is a simple and efficient numerical discretization method for the approximation of fractional-order filter (Chen 2003, 2008a, b; Chen and Vinagre 2003). The method not only can accurately approximate the fractional-order filter in time domain but also fit the frequency response very well in the low frequency band in the frequency domain (Li et al. 2010a). Figures 4.1 and 4.2 show the impulse responses and the frequency response of the approximated discrete-time IIR filter and the continuous-time fractional-order filter when $a = 0.6, b = 1$ and $T_s = 0.001\,\mathrm{s}$, respectively. The transfer function of the approximated IIR filter is

$$\frac{0.00167 - 0.006112z^{-1} + 0.008409z^{-2} - 0.005208z^{-3} + 0.00129z^{-4} - 4.785 \cdot 10^{-5}z^{-5}}{1 - 4.488z^{-1} + 8.004z^{-2} - 7.082z^{-3} + 3.104z^{-4} - 0.5383z^{-5}}.$$

(4.10)

For frequency response, the impulse response invariant discretization method works well for the band-limited (1–100 Hz) continuous-time fractional-order filters. This figure is plotted by the MATLAB code (Sheng 2010), where we used the MATLAB command [sr] = irid_doi(0.001,0.6, 1,5, 5).

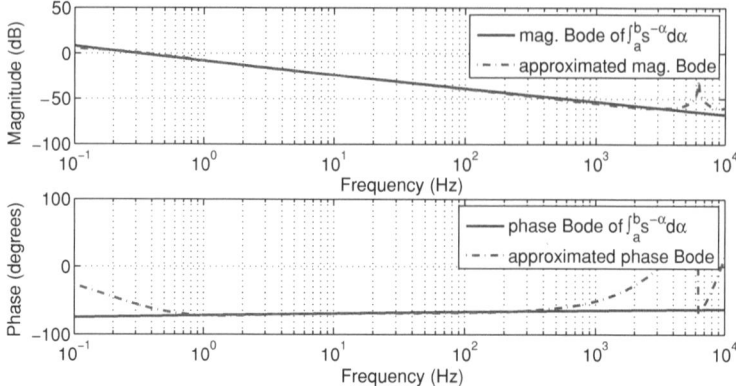

Fig. 4.2 The frequency response of the approximate discrete-time IIR filter and the continuous-time distributed-order filter when $a = 0.6$, $b = 1$ and $T_s = 0.001$ s

Remark 4.1 The algorithm proposed in Steiglitz and McBride (1965) permits more accurate identification when the impulse response is slowly varying. Therefore, it follows from Theorem 4.1 that the performance of "*stmcb*", an algorithm for finding an IIR filter with a prescribed time domain response given an input signal, in MATLAB is related to a and b. Particularly, when $0 \leq a < b \leq 1$, the approximated results are more accurate for the case when a, b are closer to 1.

It follows from Remark 4.1 that the approximated results obtained by the application of (4.7) and the discretization method have relatively good performances for $0.5 \leq a < b \leq 1$ in both time and frequency domains. Based on Theorem 4.2, and in order to extend a and b to the whole real domain, we can use the following property.

Property 4.2 *When* $0 \leq a < b \leq 0.5$, *it follows from (A) in Property 4.1 that* $\int_a^b \frac{1}{s^\alpha} d\alpha = s^{0.5-a} \int_{0.5}^{0.5+b-a} \frac{1}{s^\alpha} d\alpha$, *where* $0.5 \leq 0.5 + b - a \leq 1$.

Remark 4.2 It follows from Properties 4.1 and 4.2 that, for arbitrary $\tilde{a}, \tilde{b} \in \mathbb{R}$, $\int_{\tilde{a}}^{\tilde{b}} \frac{1}{s^\alpha} d\alpha$ can be divided into the combination of $s^\lambda (\lambda \in \mathbb{R})$ and $\int_a^b \frac{1}{s^\alpha} d\alpha$, where $a, b \in [0.5, 1]$.

Lastly, it can be shown in both time and frequency domains that the distributed-order integrator/differentiator exhibits some intermediate properties among the integer-order and fractional-order integrators/differentiators. In the frequency domain, for example, Fig. 4.3 presents the frequency responses of distributed-order integrator $\frac{1}{1-0.6} \int_{0.6}^{1} s^{-\alpha} d\alpha$, integer-order integrator $\frac{1}{s}$, and fractional-order integrators $\frac{1}{s^{0.6}}$ and $\frac{1}{s^{0.8479}}$. The fractional integrator $\frac{1}{s^{0.8479}}$ was constructed by searching the best fit to the magnitude of the distributed-order integrator $\frac{1}{1-0.6} \int_{0.6}^{1} s^{-\alpha} d\alpha$. It can be seen that the magnitude and phase of the frequency response of the distributed-order integrator are totally different from that of the

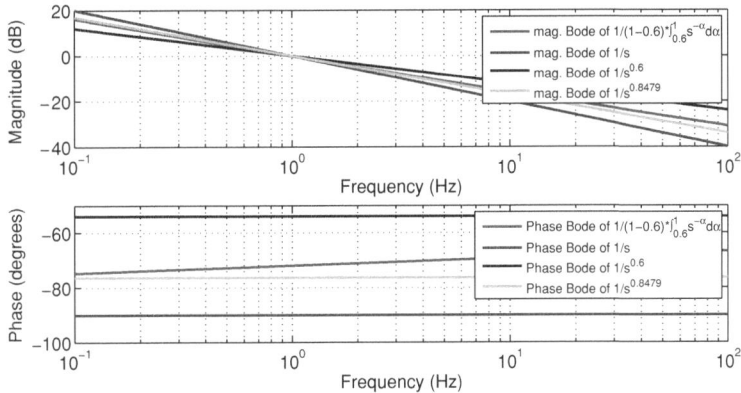

Fig. 4.3 Frequency response comparisons

fractional-order integrator and integer-order integrator. The phase of the distributed-order integrator is no longer a constant. The comparison study of these three types of integrators indicates that the distributed-order integrator exhibits distinctive frequency response characteristics. There does not exist the so-called "mean order" equivalent constant-order integrator/differentiator for the distributed-order one.

4.1.2 Distributed-Order Low-Pass Filter

In this section, we focus on the discussions of the uniformly weighted distributed-order low-pass filter

$$\frac{\lambda^{a+b} \ln \lambda}{\lambda^b - \lambda^a} \int_a^b \frac{1}{(s+\lambda)^\alpha} d\alpha, \qquad (4.11)$$

where $\lambda \geq 0$, $a < b$ are arbitrary real numbers and $\frac{\lambda^{a+b} \ln \lambda}{\lambda^b - \lambda^a}$ is the normalizing constant, such that the filter (4.11) has a unity DC gain.[2]

The classical first order low-pass filter can be rewritten as

$$\frac{1}{Ts+1} = \int_{-\infty}^{\infty} \delta(\alpha - 1) \frac{1}{(Ts+1)^\alpha} d\alpha, \qquad (4.12)$$

where $\delta(\cdot)$ denotes the Dirac–Delta function and $\frac{1}{(Ts+1)^\alpha}$ is a fractional-order low-pass filter with order $\alpha \in \mathbb{R}$.

To enable the applications of the distributed-order low-pass filter in engineering, the numerical discretization method should be applied so that the filter can be used

---

[2] When $s = 0$, DC gain of $\int_a^b \frac{1}{(s+\lambda)^\alpha} d\alpha = \int_a^b \frac{1}{\lambda^\alpha} d\alpha = \frac{1}{\ln \lambda} \left(\frac{1}{\lambda^a} - \frac{1}{\lambda^b} \right)$. So, unity gain requires gain scaling factor $\frac{\lambda^{a+b} \ln \lambda}{\lambda^b - \lambda^a}$.

in signal modeling, filter design and nonlinear system identification (Hartley and Lorenzo 2003; Adams et al. 2008; Li et al. 2010b). Let us first derive the analytical form of the filter's impulse response.

4.1.2.1 Impulse Response of the Distributed-Order Low-Pass Filter

In this section the analytical form of

$$\mathcal{L}^{-1}\left\{ \int_a^b \frac{1}{(s+\lambda)^\alpha} \mathrm{d}\alpha \right\} \tag{4.13}$$

is derived and is in a computable form in MATLAB. This will be used in the impulse response invariant discretization in the next section.

It follows from the properties of inverse Laplace transform that

$$\mathcal{L}^{-1}\left\{ \int_a^b \frac{1}{(s+\lambda)^\alpha} \mathrm{d}\alpha \right\} = e^{-\lambda t} \mathcal{L}^{-1}\left\{ \int_a^b \frac{1}{(s^\alpha)} \mathrm{d}\alpha \right\}. \tag{4.14}$$

It has been provided that by substituting $s = -xe^{-i\pi}$ and $s = xe^{i\pi}$, where $x \in (0, +\infty)$, we have, for an arbitrary $\sigma > 0$ and $b \le 1$,

$$\mathcal{L}^{-1}\left\{ \int_a^b \frac{1}{s^\alpha} \mathrm{d}\alpha \right\} = \frac{1}{\pi} \int_0^\infty \frac{e^{-xt}}{(\ln(x))^2 + \pi^2} \Big[x^{-a}(\sin(a\pi)\ln(x) + \pi\cos(a\pi))$$
$$- x^{-b}(\sin(b\pi)\ln(x) + \pi\cos(b\pi)) \Big] \mathrm{d}x. \tag{4.15}$$

Theorem 4.3 *For any $a, b \in \mathbb{R}$, we have*

$$\left| \mathcal{L}^{-1}\left\{ \int_a^b \frac{1}{(s+\lambda)^\alpha} \mathrm{d}\alpha \right\} \right| \le \frac{e^{-\lambda t}}{\pi^2} \left(\frac{M_1 t^{a-1}}{|a-1|} + \frac{M_2 t^{b-1}}{|b-1|} \right),$$

where M_1 and M_2 are finite positive constants.

Proof By using (4.15), it can be easily proved that

$$\left| \mathcal{L}^{-1}\left\{ \int_a^b \frac{1}{(s+\lambda)^\alpha} \mathrm{d}\alpha \right\} \right| \le \frac{e^{-\lambda t}}{\pi^2} \int_0^\infty e^{-xt}(x^{-a} + x^{-b}) \mathrm{d}x$$
$$= \frac{e^{-\lambda t}}{\pi^2} \left(\frac{M_1 t^{a-1}}{|a-1|} + \frac{M_2 t^{b-1}}{|b-1|} \right),$$

Fig. 4.4 The impulse response of $\frac{1}{0.4}\int_{0.6}^{1}\frac{1}{(s+1)^{\alpha}}d\alpha$

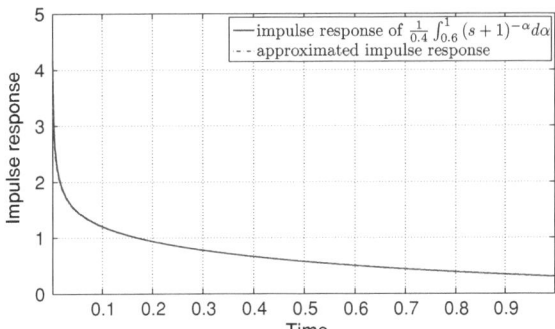

where $M_1=\int_0^\infty e^{-\tau^{1/(1-b)}}d\tau$ and $M_2=\int_0^\infty e^{-\tau^{1/(1-b)}}d\tau$ are finite positive constants for any $a, b \in \mathbb{R} \setminus \{1\}$. When $a = 1$ or $b = 1$, it is obvious that $\left|\mathcal{L}^{-1}\left\{\int_a^b \frac{1}{(s+\lambda)^\alpha}d\alpha\right\}\right| \leq +\infty$.

4.1.3 Impulse Response Invariant Discretization of DO-LPF

Now, let us consider how to discretize the $G(s)$ given sampling period T_s. Our goal is to get a discretized version of $G(s)$, denoted by $G_d(z^{-1})$ with a constraint that $G_d(z^{-1})$ and $G(s)$ have the same impulse responses. Since the analytical impulse response of $G(s)$ had already been derived in Sect. 4.1.2.1, it is relatively straightforward to obtain the impulse response invariant discretized version of $G(s)$ via the well-known Prony technique (Chen 2003, 2008a, b; Chen and Vinagre 2003). In other words, the discretization impulse response can be obtained by using the continuous time impulse response as follows:

$$g(n) = T_s g(n T_s),\qquad(4.16)$$

where $n = 0, 1, 2, \cdots$ and T_s is the sampling period.

Figure 4.4 shows the magnitude and phase of the frequency response of the approximate discrete-time IIR filter and the continuous-time fractional distributed order filter $\frac{1}{0.4}\int_{0.6}^{1}\frac{1}{(s+1)^\alpha}d\alpha$. The transfer function of the approximate IIR filter $H(z)$ is

$$\frac{0.00417 - 0.01509z^{-1} + 0.02048z^{-2} - 0.01248z^{-3} + 0.003019z^{-4} - 0.0001022z^{-5}}{1 - 4.445z^{-1} + 7.844z^{-2} - 6.859z^{-3} + 2.967z^{-4} - 0.5066z^{-5}}.$$

For frequency responses, the impulse response invariant discretization method works well for the continuous-time fractional-order filters. The continuous and discretized

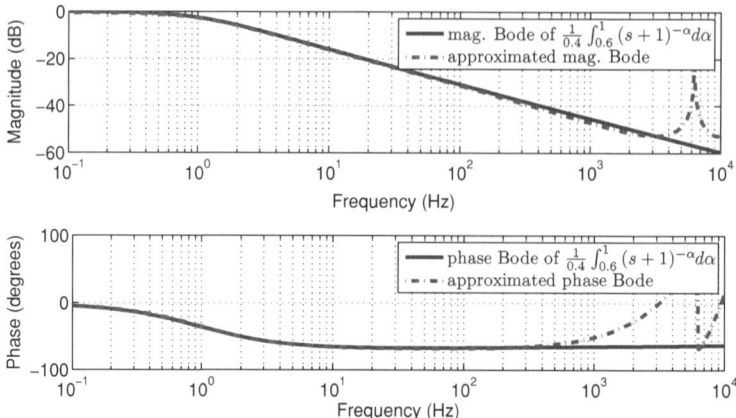

Fig. 4.5 The frequency response of $\frac{1}{0.4}\int_{0.6}^{1}\frac{1}{(s+1)^{\alpha}}d\alpha$

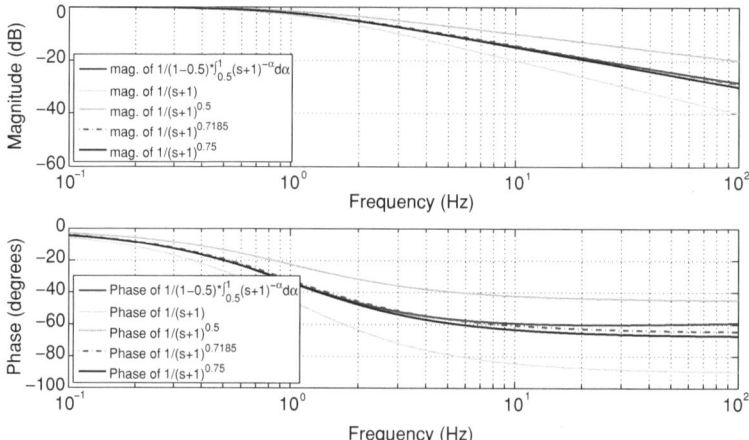

Fig. 4.6 The comparisons of distributed-order low-pass filter with several integer-order and constant-order low-pass filters

impulse response and frequency response are also shown in Figs. 4.4 and 4.5, where $T_s = 0.001$ s. Then, several low-pass filters are compared and shown in Fig. 4.6.

It can be seen that the distributed-order low-pass filter is an intermediate one among integer-order and fractional-order low-pass filters.

From this section, we can see that, distributed order operator can be fruitfully used for distributed-order filters. The next application example is closer to engineering setting on distributed-order damping.

4.2 Application II: Optimal Distributed-Order Damping

A damper is a valuable component for reducing the amplitude of dynamic instabil-
ities or resonances in system stabilization (Shafieezadeh et al. 2008). In physics
and engineering, the mathematical model of the conventional damping can be
represented by

$$f(t) = -cv(t) = -c\frac{dx(t)}{dt}, \tag{4.17}$$

where $f(t)$ is the time varying force, c is the viscous damping coefficient, $v(t)$
is the velocity, and $x(t)$ is the displacement (Komkov 1972). Taking advantage of
fractional calculus, fractional-order damping with a viscoelastic damping element
provides a better model to describe a damping system (Koeller 1984). Fractional-
order damping is modeled as a force proportional to the fractional-order derivative
of the displacement (Lion 1997)

$$f(t) = c_0 D_t^\alpha x(t), \tag{4.18}$$

where $_0D_t^\alpha x(t)$ is the fractional-order derivative defined in Chap. 1 of Podlubny
(1999). Motivated by potential benefits of fractional damping, many efforts have been
made to investigate the modeling of systems with damping materials using fractional-
order differential operators (Rossikhin and Shitikova 1997; Padovan and Guo 1988;
Shokooh 1999; Rüdinger 2006; De Espíndola et al. 2008; Dalir and Bashour 2010).
However, up to now, little attention has been paid to time-delayed fractional-order
damping, and distributed-order fractional damping. In this section, we investigate the
potential benefits of a non-delayed fractional-order damping system, a time-delayed
fractional-order damping system, and a distributed-order fractional damping system.

In order to design an optimal transfer function form, the performance of a control
system should be measured, and the parameters of the system should be adjusted to
deliver the desirable response. The performance of a system is usually specified by
several time response indices for a step input, such as rise time, peak time, overshoot,
and so on (Dorf 1989). Furthermore, the performance index, a scalar, is adequately
used to represent the important system specifications instead of a set of indices. The
transfer function of a system is considered as an optimal form when the system
parameters are adjusted so that the performance index reaches an extremum value
(Dorf 1989). The well-known integral performance indices are the integral of absolute
error (IAE), the integral of squared error (ISE), the integral of time multiplied absolute
error (ITAE), the integral of time multiplied squared error (ITSE), and the integral
of squared of time multiplied error (ISTE) (Dorf 1989; Tavazoei 2010). Hartley and
Lorenzo studied the single term damper that minimizes the time domain ISE and
ITSE, and found that the optimal fractional-order damping is more optimal than
the optimal integer-order damping (Hartley and Lorenzo 2004). In this section, we
investigate three types of optimal fractional-order damping systems using frequency-
domain optimization method. In frequency domain, the time-delayed fractional-order

and distributed-order fractional damping systems are optimized using ISE criterion. The comparisons of an optimal integer-order damping system and three optimal fractional-order damping systems indicate that ISE optimal fractional-order damping systems perform better than ISE optimal integer-order damping systems. The optimal time-delayed fractional-order damping system performs the best among the optimal integer-order damping system and optimal fractional-order damping systems.

An interesting fact revealed in this section is that, the time delay, which is usually regarded as a detrimental negative destabilizing factor, can sometimes be used to gain benefit in control systems. And the distributed-order fractional damper performs as well as fractional-order damping. Furthermore, the distributed-order fractional damper has great potential to improve the damping by choosing the appropriate order weighting function as the order-dependent viscoelastic damping coefficient.

4.2.1 Distributed-Order Damping in Mass-Spring Viscoelastic Damper System

In this section, we explain the distributed-order fractional mass-spring viscoelastic damper system in detail. This can serve as a physical interpretation of the origin and need of distributed operator.

At first, we briefly review the mass-spring-damper, mass-spring viscoelastic damper, and time-delayed mass-spring viscoelastic damper. An ideal mass-spring-damper system with mass m, spring constant k, and viscous damper of damping coefficient c can be described by a second-order differential equation

$$f(t) = m\frac{d^2x(t)}{dt^2} + c\frac{dx(t)}{dt} + kx(t), \tag{4.19}$$

where $f(t)$ is the time varying force on the mass, $x(t)$ is the displacement of the mass relative to a fixed point of reference. The transfer function from force to displacement for the ideal mass-spring-damper system can be expressed as

$$G(s) = \frac{1}{ms^2 + cs + k}. \tag{4.20}$$

A mass-spring viscoelastic damper system can be described by a fractional-order differential equation

$$f(t) = m\frac{d^2x(t)}{dt^2} + c\,_0D_t^\alpha x(t) + kx(t), \tag{4.21}$$

where $0 < \alpha < 2$. The transfer function form of a mass-spring viscoelastic damper system can be expressed as

Fig. 4.7 A distributed-order fractional mass-spring viscoelastic damper system

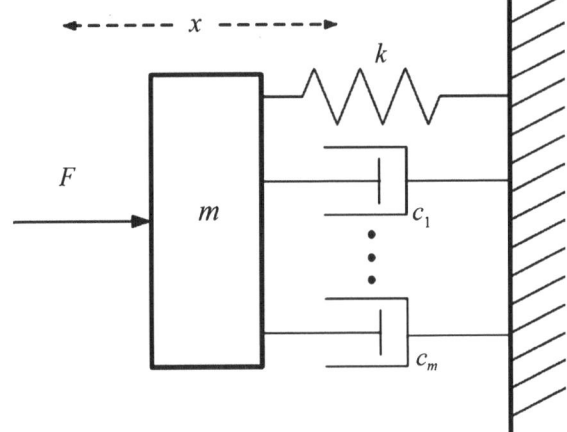

$$G(s) = \frac{1}{ms^2 + cs^\alpha + k}. \tag{4.22}$$

Similarly, the transfer function form of a time-delayed mass-spring viscoelastic damper system can be expressed as

$$G(s) = \frac{1}{ms^2 + cs^\alpha e^{-\tau s} + k}, \tag{4.23}$$

where $0 < \alpha < 2$.

A distributed-order fractional mass-spring viscoelastic damper system with mass m (in kilograms), spring constant k (in Newton per meter) and an assembly of viscoelastic dampers of damping coefficient c_i $(1 < i < n)$ is subject to the spring force

$$f_s(t) = -kx(t), \tag{4.24}$$

and damping force

$$f_d(t) = -\sum_{i=1}^{n} c_i \cdot {}_0D_t^{\alpha_i} x(t), \tag{4.25}$$

where c_i is the viscoelastic damping coefficient. Figure 4.7 illustrates a distributed-order fractional mass-spring viscoelastic damper system. According to the Newton's second law, the total force $f_{tot}(t)$ on the body is

$$f_{tot}(t) = ma = m\frac{d^2 x(t)}{dt^2}, \tag{4.26}$$

where a is the acceleration (in meters per second squared) of the mass, and $x(t)$ is the displacement (in meters) of the mass relative to a fixed point of reference. The time varying force on the mass can be represented by

$$f(t) = f_{tot}(t) - f_d(t) - f_s(t)$$

$$= m\frac{d^2x(t)}{dt^2} + \sum_{i=1}^{n} c_i \, _0D_t^{\alpha_i} x(t) + kx(t). \tag{4.27}$$

Assuming elements with orders that vary from a to b, the above mass-spring viscoelastic damper system of (4.27) can be replaced by an integral over the system order,

$$f(t) = m\frac{d^2x(t)}{dt^2} + \int_a^b c(\alpha) \, _0D_t^\alpha x(t)d\alpha + kx(t), \tag{4.28}$$

where $0 < a < b < 2$. The transfer function from the force to displacement x for the spring-mass-viscoelastic damper system of (4.28) can be expressed as

$$G(s) = \frac{X(s)}{F(s)} = \frac{1}{ms^2 + \int_a^b c(\alpha)s^\alpha d\alpha + k}. \tag{4.29}$$

What we are concentrating on in this study is the normalized transfer functions of above three types of the spring-mass-viscoelastic damper systems. They are: the normalized transfer function of the spring-mass-viscoelastic damper system

$$G(s) = \frac{1}{s^2 + cs^\alpha + 1}, \quad 0 < \alpha < 2; \tag{4.30}$$

the normalized transfer function of the time-delayed spring-mass-viscoelastic damper system

$$G(s) = \frac{1}{s^2 + cs^\alpha e^{-\tau s} + 1}, \quad 0 < \alpha < 2, \tag{4.31}$$

and the normalized transfer function of the constant damper coefficient distributed-order spring-mass-viscoelastic damper system

$$G(s) = \frac{1}{s^2 + c\int_a^b s^\alpha d\alpha + 1}, \quad 0 < a < b < 2. \tag{4.32}$$

4.2.2 Frequency-Domain Method Based Optimal Fractional-Order Damping Systems

The ISE optimal integer-order damping system with transfer function

$$G(s) = \frac{1}{s^2 + s + 1} \tag{4.33}$$

has been investigated in Ogata (1970), and the ISE optimal fractional-order damping
with transfer function

$$G(s) = \frac{1}{s^2 + 0.8791s^{0.8459} + 1} \tag{4.34}$$

has been found in Lorenzo and Hartley(2002) using a frequency-domain method. In
this section, ISE optimal time-delayed and distributed-order fractional mass-spring
viscoelastic damper systems are studied in frequency-domain. The ISE performance
measure is the integral of the squared error of the step response $e(t) = u(t) - x(t)$

$$J_{ISE} = \int_0^\infty e^2(t)dt, \tag{4.35}$$

where $x(t)$ is the output of the system (D'Azzo et al. 2003). Using Parseval's identity

$$J_{ISE} = \int_0^\infty e^2(t)dt = \frac{1}{2\pi} \int_{-\infty}^\infty |E(j\omega)|^2 d\omega, \tag{4.36}$$

where $E(j\omega)$ is the Fourier transform of the error $e(t)$. For a system with transfer
function $G(s)$, the Laplace transform of the error can be written as

$$E(s) = \frac{1}{s} - \frac{1}{s}G(s). \tag{4.37}$$

In frequency domain, (4.37) is represented by

$$E(j\omega) = \frac{1}{j\omega} - \frac{1}{j\omega}G(j\omega). \tag{4.38}$$

For a time-delayed spring-mass-viscoelastic damper system with the normalized
transfer function (4.31), the Laplace transform of the step response error is

$$E(s) = \frac{1}{s} - \frac{1}{s}\left(\frac{1}{s^2 + cs^\alpha e^{-\tau s} + 1}\right) = \frac{1}{s}\left(\frac{s^2 + cs^\alpha e^{-\tau s}}{s^2 + cs^\alpha e^{-\tau s} + 1}\right). \tag{4.39}$$

The frequency response of the error is

$$E(j\omega) = \frac{1}{j\omega}\left(\frac{(j\omega)^2 + c(j\omega)^\alpha e^{-\tau(j\omega)}}{(j\omega)^2 + c(j\omega)^\alpha e^{-\tau(j\omega)} + 1}\right). \tag{4.40}$$

Using the frequency-domain method in Hartley and Lorenzo (2004), the minimum
$J_{ISE} = 0.8102$ was obtained when $\tau = 0.635$, $c = 1.12$ and $\alpha = 1.05$. The step
response using optimum coefficients for the ISE criterion is given in Fig. 4.8.

For a mass-spring viscoelastic damper model with the normalized distributed-
order fractional transfer function (4.32), the Laplace transform of the step response

Fig. 4.8 Step responses of
ISE optimal damping systems
based on frequency-domain
method

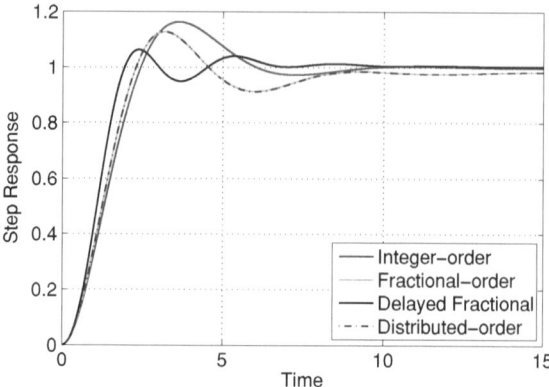

error is

$$E(s) = \frac{1}{s} - \frac{1}{s}\left(\frac{1}{s^2 + c\int_a^b s^\alpha d\alpha + 1}\right)$$

$$= \frac{1}{s}\left(\frac{\ln(s)s^2 + c\left(s^b - s^a\right)}{\ln(s)s^2 + c\left(s^b - s^a\right) + \ln(s)}\right). \tag{4.41}$$

The frequency response of the error is

$$E(j\omega) = \frac{1}{j\omega}\left(\frac{\ln(j\omega)(j\omega)^2 + c[(j\omega)^b - (j\omega)^a]}{\ln(j\omega)(j\omega)^2 + c[(j\omega)^b - (j\omega)^a] + \ln(j\omega)}\right). \tag{4.42}$$

Then, we can search the optimum coefficients of the distributed-order fractional
damping system. The optimum coefficients are $a = 0.8015$, $b = 0.8893$ and $c = 10$,
which can minimize the ISE performance measure to $J_{ISE} = 0.9494$. Figure 4.8
shows the step responses of integer-order, non-delayed fractional-order, time-delayed
fractional-order, and distributed-order fractional damping systems using optimum
coefficients for ISE. It can be seen that the step responses of optimal distributed-
order fractional damping system with transfer function (4.32) are almost as good as
that of the optimal non-delayed fractional-order damping system. The optimal time-
delayed fractional-order damping system performs the best among these four types
of damping systems. The ISE optimal forms and ISE performance indexes of integer-
order, non-delayed fractional-order, time-delayed fractional order, and distributed-
order fractional damping systems are summarized in Table 4.1.

In this section, we tried to determine the optimal non-delayed fractional-order
damping, time-delayed fractional-order damping, and optimal distributed order
fractional damping based on ISE performance criterion. The comparisons of the
step responses of the integer-order and the three types of fractional-order damping
systems indicate that the optimal fractional-order damping systems achieve much

Table 4.1 ISE optimum coefficients and minimum ISE performance indexes using frequency-domain method

	Optimal form	J_{ISE}
Integer	$G_{ISE}(s) = 1/(s^2 + s + 1)$	1.0000
Fractional	$G_{ISE}(s) = 1/(s^2 + 0.8791s^{0.8459} + 1)$	0.9494
Delayed	$G_{ISE}(s) = 1/(s^2 + 1.12s^{1.05}e^{-0.635s} + 1)$	0.8102
Distributed	$G_{ISE}(s) = 1/(s^2 + 10\int_{0.8015}^{0.8893} s^\alpha d\alpha + 1)$	0.9494

better step responses than optimal integer-order systems in some instances, but sometimes the integer-order damping systems performs as well as fractional-order ones. Furthermore, time delay can sometimes be used to gain benefit in control systems, and, especially, the fractional-order damping plus properly chosen delay can bring outstanding performance. Time-delayed fractional-order damping systems can produce a faster rise time and less overshoot than others.

4.3 Chapter Summary

This chapter shows two generic application examples using distributed-order operator: distributed-order signal processing and optimal distributed-order damping. In distributed-order signal processing, the simplest case of distributed-order integrator/differentiator is discussed first followed by the discussion of distributed-order low-pass filter. Specifically, we derived the impulse response functions of the distributed-order integrator/differentiator and fractional-order distributed low-pass filter from the complex path integral expressed in the definite integral form. Based on these results, we obtained some asymptotic properties of the impulse responses, and we can accurately compute the integrals on the whole time domain. Moreover, for practical applications, we presented a technique known as "impulse-response-invariant discretization" to perform the discretization of these two distributed-order filters. We are able to show that the distributed-order fractional filters have some unique features compared with the classical integer-order or constant-order fractional filters.

Then, optimal distributed-order damping strategies are given for a given standard form of second order system known as distributed-order fractional mass-spring viscoelastic damper system. Frequency-domain method based optimal fractional-order damping systems are numerically solved. Although the distributed-order fractional damping system with uniform weights does not perform obviously better than non-delayed and time-delayed fractional-order damping systems, it is believed to have much potential to improve the damping system by choosing an appropriate viscoelastic damping coefficient weighting function. So, our next step is to explore the benefits of distributed-order fractional damping system with an "optimal" viscoelastic coefficient weighting function.

References

Adams JL, Hartley TT, Lorenzo CF (2008) Identification of complex order-distributions. J Vib Control 14(9–10):1375–1388

Barbosa RS, Machado JAT (2006) Implementation of discrete-time fractional-order controllers based on LS approximations. Acta Polytech Hung 3(4):5–22

Caputo M (1969) Elasticità e dissipazione. Zanichelli, Bologna

Chen YQ (2003) Low-pass IIR digital differentiator design. http://www.mathworks.com/matlabcentral/fileexchange/3517

Chen YQ (2008a) Impulse response invariant discretization of fractional order integrators or differentiators. http://www.mathworks.com/matlabcentral/fileexchange/21342

Chen YQ (2008b) Impulse response invariant discretization of fractional order low-pass filters. http://www.mathworks.com/matlabcentral/fileexchange/21365

Chen YQ, Moore KL (2002) Discretization schemes for fractional-order differentiators and integrators. IEEE Trans Circuits Syst I Fundam Theory Appl 49(3):363–367

Chen YQ, Vinagre BM (2003) A new IIR-type digital fractional order differentiator. Signal Process 83(11):2359–2365

Chen YQ, Vinagre BM, Podlubny I (2004) Continued fraction expansion approaches to discretizing fractional order derivatives—an expository review. Nonlinear Dyn 38(16):155–170, Dec 2004

Dalir M, Bashour M (2010) Applications of fractional calculus. Appl Math Sci 4(21–24):1021–1032

D'Azzo JJ, Houpis CH, Sheldon SN (2003) Linear control system analysis and design, 5th edn. CRC Press, Boca Raton

De Espíndola JJ, Bavastri CA, De Oliveira Lopes EM (2008) Design of optimum systems of viscoelastic vibration absorbers for a given material based on the fractional calculus model. J Vib Control 14(9–10):1607–1630

Dorf RC (1989) Modern control systems. Addison-Wesley Longman, Boston

Ferdi Y (2009) Impulse invariance-based method for the computation of fractional integral of order $0 < \alpha < 1$. Comput Electr Eng 35(5):722–729

Hartley TT, Lorenzo CF (2003) Fractional-order system identification based on continuous order-distributions. Signal Process 83(11):2287–2300

Hartley TT, Lorenzo CF (2004) A frequency-domain approach to optimal fractional-order damping. Nonlinear Dyn 38(1–2):69–84

Koeller RC (1984) Applications of fractional calculus to the theory of viscoelasticity. J Appl Mech 51(2):299–307

Komkov V (1972) Optimal control theory for the damping of vibrations of simple elastic systems. Springer, Berlin

Li Y, Sheng H, Chen YQ (2010a) Impulse response invariant discretization of distributed order low-pass filter. http://www.mathworks.com/matlabcentral/fileexchange/authors/82211

Li Y, Sheng H, Chen YQ (2010b) On distributed order integrator/differentiator. Signal Process 91(5):1079–1084

Li Y, Sheng H, Chen YQ (2011) Analytical impulse response of a fractional second order filter and its impulse response invariant discretization. Signal Process 91(3):498–507

Lion A (1997) On the thermodynamics of fractional damping elements. Continuum Mech Thermodyn 9(2):83–96

Lorenzo CF, Hartley TT (2002) Variable order and distributed order fractional operators. Nonlinear Dyn 29(1–4):57–98

Lubich C (1986) Discretized fractional calculus. SIAM J Math Anal 17(3):704–719

Machado JAT (1997) Analysis and design of fractional-order digital control systems. Syst Anal Model Simul 27(2–3):107–122

Ogata K (1970) Modern control engineering. Prentice-Hall, Upper Saddle River

Oustaloup A (1981) Fractional order sinusoidal oscillators: optimization and their use in highly linear FM modulation. IEEE Trans Circuits Syst 28(10):1007–1009

Oustaloup A, Levron F, Mathieu B, Nanot FM (2000) Frequency-band complex noninteger differentiator: characterization and synthesis. IEEE Trans Circuits Syst I Fundam Theory Appl 47(1): 25–39

Padovan J, Guo YH (1988) General response of viscoelastic systems modelled by fractional operators. J Frankl Inst 325(2):247–275

Podlubny I (1999) Fractional differential equations. Academic Press, San Diego

Radwan AG, Soliman AM, Elwakil AS (2007) Design equations for fractional-order sinusoidal oscillators: practical circuit examples. In: Proceedings of the internatonal conference on microelectronics, pp 89–92, Dec 2007

Radwan AG, Elwakil AS, Soliman AM (2008) Fractional-order sinusoidal oscillators: design procedure and practical examples. IEEE Trans Circuits Syst I Regul Pap 55(7):2051–2063

Rossikhin YA, Shitikova MV (1997) Application of fractional derivatives to the analysis of damped vibrations of viscoelastic single mass systems. Acta Mech 120(1–4):109–125

Rüdinger F (2006) Tuned mass damper with fractional derivative damping. Eng Struct 28(13):1774–1779

Sheng H (2010) Impulse response invariant discretization of distributed order integrator. http://www.mathworks.com/matlabcentral/fileexchange/26380

Shafieezadeh A, Ryan K, Chen YQ (2008) Fractional order filter enhanced LQR for seismic protection of civil structures. J Comput Nonlinear Dyn 3(2):020201.1–021404.7

Shokooh A (1999) A comparison of numerical methods applied to a fractional model of damping materials. J Vib Control 5(3):331–354

Steiglitz K, McBride L (1965) A technique for the identification of linear systems. IEEE Trans Autom Control 10(4):461–464

Tavazoei MS (2010) Notes on integral performance indices in fractional-order control systems. J Process Control 20(3):285–291

Vinagre BM, Chen YQ, Petras I (2003) Two direct Tustin discretization methods for fractional-order differentiator/integrator. J Frankl Inst 340(5):349–362

Chapter 5
Numerical Solution of Differential Equations of Distributed Order

5.1 Introduction

In this chapter we present a general approach to numerical solution to discretization of distributed-order derivatives and integrals, and to numerical solution of ordinary and partial differential equations of distributed order.

This approach is based on the matrix form representation of discretized fractional operators of constant order introduced for the first time in Podlubny (2000) and extended further in the works (Podlubny et al. 2009a, b; Skovranek et al. 2010; Podlubny et al. 2011).

This approach unifies the numerical differentiation of arbitrary (including integer) order and the n-fold integration, using the so-called triangular matrices. Applied to numerical solution of differential equations, it also unifies the solution of integer- and fractional-order partial differential equations. The matrix approach lead to significant simplification of the numerical solution of partial differential equations as well, and it is general enough to deal with different types of partial fractional differential equations.

In this chapter we extend the range of applicability of the matrix approach to discretization of distributed-order derivatives and integrals, and to numerical solution of distributed-order differential equations (both ordinary and partial).

Since the distributed-order operators are represented by integrals of weighted constant-order operators, we necessarily first introduce the matrix approach to discretization of constant order and then demonstrate how this method can be extended to allow numerical solution of distributed-order differential equations.

5.2 Triangular Strip Matrices

We use matrices of a specific structure, which are called *triangular strip matrices* (Podlubny 2000; Suprunenko and Tyshkevich 1966), and which have been also

mentioned in Bulgakov (1954) and Gantmakher (1988). We will need lower triangular strip matrices,

$$
L_N = \begin{bmatrix}
\omega_0 & 0 & 0 & 0 & \cdots & 0 \\
\omega_1 & \omega_0 & 0 & 0 & \cdots & 0 \\
\omega_2 & \omega_1 & \omega_0 & 0 & \cdots & 0 \\
\ddots & \ddots & \ddots & \ddots & \cdots\cdots & \\
\omega_{N-1} & \ddots & \omega_2 & \omega_1 & \omega_0 & 0 \\
\omega_N & \omega_{N-1} & \ddots & \omega_2 & \omega_1 & \omega_0
\end{bmatrix},
\tag{5.1}
$$

and upper triangular strip matrices,

$$
U_N = \begin{bmatrix}
\omega_0 & \omega_1 & \omega_2 & \ddots & \omega_{N-1} & \omega_N \\
0 & \omega_0 & \omega_1 & \ddots & \ddots & \omega_{N-1} \\
0 & 0 & \omega_0 & \ddots & \omega_2 & \ddots \\
0 & 0 & 0 & \ddots & \omega_1 & \omega_2 \\
\cdots\cdots\cdots & & & \cdots & \omega_0 & \omega_1 \\
0 & 0 & 0 & \cdots & 0 & \omega_0
\end{bmatrix}.
\tag{5.2}
$$

A lower (upper) triangular strip matrix is completely described by its first column
(row). Therefore, if we define the truncation operation, $\mathrm{trunc}_N\,(\cdot)$, which truncates
(in a general case) the power series $\rho(z)$,

$$
\rho(z) = \sum_{k=0}^{\infty} \omega_k z^k
\tag{5.3}
$$

to the polynomial $\rho_N(z)$,

$$
\mathrm{trunc}_N\,(\rho(z)) \stackrel{\mathrm{def}}{=} \sum_{k=0}^{N} \omega_k z^k = \rho_N(z),
\tag{5.4}
$$

then we can consider the function $\rho(z)$ as a generating series for the set of lower
(or upper) triangular matrices L_N (or U_N), $N = 1, 2, \ldots$

It was shown in Podlubny (2000) that operations with triangular strip matrices,
such as addition, subtraction, multiplication, and inversion, can be expressed in the
form of operations with their generating series (5.3).

Among properties of triangular strip matrices it should be noticed that if matrices
C and D are both lower (upper) triangular strip matrices, then they commute:

$$
CD = DC.
\tag{5.5}
$$

5.3 Kronecker Matrix Product

The Kronecker product $A \otimes B$ of the $n \times m$ matrix A and the $p \times q$ matrix B

$$
A = \begin{bmatrix} a_{11} & a_{12} & \dots & a_{1m} \\ a_{21} & a_{22} & \dots & a_{2m} \\ \vdots & \vdots & \ddots & \vdots \\ a_{n1} & a_{n2} & \dots & a_{nm} \end{bmatrix}, \qquad B = \begin{bmatrix} b_{11} & b_{12} & \dots & b_{1q} \\ b_{21} & b_{22} & \dots & b_{2q} \\ \vdots & \vdots & \ddots & \vdots \\ b_{p1} & b_{p2} & \dots & b_{pq} \end{bmatrix}, \tag{5.6}
$$

is the $np \times mq$ matrix having the following block structure:

$$
A \otimes B = \begin{bmatrix} a_{11}B & a_{12}B & \dots & a_{1m}B \\ a_{21}B & a_{22}B & \dots & a_{2m}B \\ \vdots & \vdots & \ddots & \vdots \\ a_{n1}B & a_{n2}B & \dots & a_{nm}B \end{bmatrix}. \tag{5.7}
$$

For example, if

$$
A = \begin{bmatrix} 1 & 2 \\ 0 & -3 \end{bmatrix}, \qquad B = \begin{bmatrix} 1 & 2 & 3 \\ 4 & 5 & 6 \end{bmatrix}, \tag{5.8}
$$

then

$$
A \otimes B = \begin{bmatrix} 1 & 2 & 3 & 2 & 4 & 6 \\ 4 & 5 & 6 & 8 & 10 & 12 \\ 0 & 0 & 0 & -3 & -6 & -9 \\ 0 & 0 & 0 & -12 & -15 & -18 \end{bmatrix}. \tag{5.9}
$$

Among many known interesting properties of the Kronecker product we would like to recall those that are important for the subsequent sections. Namely (Loan 2000),

- if A and B are band matrices, then $A \otimes B$ is also a band matrix,
- if A and B are lower (upper) triangular, then $A \otimes B$ is also lower (upper) triangular.

We will also need two specific Kronecker products, namely the products $E_n \otimes A$ and $A \otimes E_m$, where E_n is an $n \times n$ identity matrix. For example, if A is a 2×3 matrix

$$
A = \begin{bmatrix} a_{11} & a_{12} & a_{13} \\ a_{21} & a_{22} & a_{23} \end{bmatrix} \tag{5.10}
$$

then

$$
E_2 \otimes A = \begin{bmatrix} a_{11} & a_{12} & a_{13} & 0 & 0 & 0 \\ a_{21} & a_{22} & a_{23} & 0 & 0 & 0 \\ 0 & 0 & 0 & a_{11} & a_{12} & a_{13} \\ 0 & 0 & 0 & a_{21} & a_{22} & a_{23} \end{bmatrix}, \tag{5.11}
$$

$$A \otimes E_3 = \begin{bmatrix} a_{11} & 0 & 0 & a_{12} & 0 & 0 & a_{13} & 0 & 0 \\ 0 & a_{11} & 0 & 0 & a_{12} & 0 & 0 & a_{13} & 0 \\ 0 & 0 & a_{11} & 0 & 0 & a_{12} & 0 & 0 & a_{13} \\ a_{21} & 0 & 0 & a_{22} & 0 & 0 & a_{23} & 0 & 0 \\ 0 & a_{21} & 0 & 0 & a_{22} & 0 & 0 & a_{23} & 0 \\ 0 & 0 & a_{21} & 0 & 0 & a_{22} & 0 & 0 & a_{23} \end{bmatrix}. \tag{5.12}$$

This illustrates that left multiplication of $A_{n \times m}$ by E_n creates an $n \times n$ block diagonal matrix by repeating the matrix A on the diagonal, and that right multiplication of $A_{n \times m}$ by E_m creates a sparse matrix made of $n \times m$ diagonal blocks.

5.4 Discretization of Ordinary Fractional Derivatives of Constant Order

It follows from Podlubny (2000), that the left-sided Riemann-Liouville or Caputo fractional derivative $v^{(\alpha)}(t) = {}_0 D_t^\alpha v(t)$ can be approximated in all nodes of the equidistant discretization net $t = j\tau$ $(j = 0, 1, \ldots, n)$ simultaneously with the help of the upper triangular strip matrix $B_n^{(\alpha)}$ as[1]

$$\begin{bmatrix} v_n^{(\alpha)} & v_{n-1}^{(\alpha)} & \cdots & v_1^{(\alpha)} & v_0^{(\alpha)} \end{bmatrix}^T = B_n^{(\alpha)} \begin{bmatrix} v_n & v_{n-1} & \cdots & v_1 & v_0 \end{bmatrix}^T \tag{5.13}$$

where

$$B_n^{(\alpha)} = \frac{1}{\tau^\alpha} \begin{bmatrix} w_0^{(\alpha)} & w_1^{(\alpha)} & \cdots & \cdots & w_{n-1}^{(\alpha)} & w_n^{(\alpha)} \\ 0 & w_0^{(\alpha)} & w_1^{(\alpha)} & \cdots & \cdots & w_{n-1}^{(\alpha)} \\ 0 & 0 & w_0^{(\alpha)} & w_1^{(\alpha)} & \cdots & \cdots \\ \cdots & \cdots & \cdots & \ddots & \ddots & \ddots \\ 0 & \cdots & 0 & 0 & w_0^{(\alpha)} & w_1^{(\alpha)} \\ 0 & 0 & \cdots & 0 & 0 & w_0^{(\alpha)} \end{bmatrix}, \tag{5.14}$$

$$w_j^{(\alpha)} = (-1)^j \binom{\alpha}{j}, \qquad j = 0, 1, \ldots, n. \tag{5.15}$$

---

[1] Here due to the use of the descending numbering of discretization nodes the roles of the matrices $B_n^{(\alpha)}$ (originally for backward fractional differences) and $F_n^{(\alpha)}$ (originally for forward fractional differences) are swapped in comparison with Podlubny (2000), where these matrices were introduced for the first time. However, we would like to preserve the notation $B_n^{(\alpha)}$ for the case of the backward fractional differences approximation and $F_n^{(\alpha)}$ for the case of the forward fractional differences approximation.

Similarly, the right-sided Riemann-Liouville or Caputo fractional derivative $v^{(\alpha)}(t) = {}_tD_b^\alpha v(t)$ can be approximated in all nodes of the equidistant discretization net $t = j\tau$ $(j = 0, 1, \ldots, n)$ simultaneously with the help of the lower triangular strip matrix $F_n^{(\alpha)}$:

$$\begin{bmatrix} v_n^{(\alpha)} & v_{n-1}^{(\alpha)} & \cdots & v_1^{(\alpha)} & v_0^{(\alpha)} \end{bmatrix}^T = F_n^{(\alpha)} \begin{bmatrix} v_n & v_{n-1} & \cdots & v_1 & v_0 \end{bmatrix}^T, \tag{5.16}$$

$$F_n^{(\alpha)} = \frac{1}{\tau^\alpha} \begin{bmatrix} \omega_0^{(\alpha)} & 0 & 0 & 0 & \cdots & 0 \\ \omega_1^{(\alpha)} & \omega_0^{(\alpha)} & 0 & 0 & \cdots & 0 \\ \omega_2^{(\alpha)} & \omega_1^{(\alpha)} & \omega_0^{(\alpha)} & 0 & \cdots & 0 \\ \ddots & \ddots & \ddots & \ddots & \cdots & \cdots \\ \omega_{n-1}^{(\alpha)} & \ddots & \omega_2^{(\alpha)} & \omega_1^{(\alpha)} & \omega_0^{(\alpha)} & 0 \\ \omega_n^{(\alpha)} & \omega_{n-1}^{(\alpha)} & \ddots & \omega_2^{(\alpha)} & \omega_1^{(\alpha)} & \omega_0^{(\alpha)} \end{bmatrix}. \tag{5.17}$$

The symmetric Riesz derivative of order β can be approximated based on its definition as a half-sum of the approximations (5.13) and (5.16) for the left- and right-sided Riemann-Liouville derivatives. We, however, prefer using the centered fractional differences approximation of the symmetric Riesz derivative suggested recently by Ortigueira (2006) and Ortigueira and Batista (2008), which gives

$$\begin{bmatrix} v_m^{(\beta)} & v_{m-1}^{(\beta)} & \cdots & v_1^{(\beta)} & v_0^{(\beta)} \end{bmatrix}^T = R_m^{(\beta)} \begin{bmatrix} v_m & v_{m-1} & \cdots & v_1 & v_0 \end{bmatrix}^T \tag{5.18}$$

with the following symmetric matrix:

$$R_m^{(\beta)} = h^{-\beta} \begin{bmatrix} \omega_0^{(\beta)} & \omega_1^{(\beta)} & \omega_2^{(\beta)} & \omega_3^{(\beta)} & \cdots & \omega_m^{(\beta)} \\ \omega_1^{(\beta)} & \omega_0^{(\beta)} & \omega_1^{(\beta)} & \omega_2^{(\beta)} & \cdots & \omega_{m-1}^{(\beta)} \\ \omega_2^{(\beta)} & \omega_1^{(\beta)} & \omega_0^{(\beta)} & \omega_1^{(\beta)} & \cdots & \omega_{m-2}^{(\beta)} \\ \ddots & \ddots & \ddots & \ddots & \cdots & \cdots \\ \omega_{m-1}^{(\beta)} & \ddots & \omega_2^{(\beta)} & \omega_1^{(\beta)} & \omega_0^{(\beta)} & \omega_1^{(\beta)} \\ \omega_m^{(\beta)} & \omega_{m-1}^{(\beta)} & \ddots & \omega_2^{(\beta)} & \omega_1^{(\beta)} & \omega_0^{(\beta)} \end{bmatrix}, \tag{5.19}$$

$$\omega_k^{(\beta)} = \frac{(-1)^k \, \Gamma(\beta + 1) \, \cos(\beta\pi/2)}{\Gamma(\beta/2 - k + 1) \, \Gamma(\beta/2 + k + 1)}, \qquad k = 0, 1, \ldots, m. \tag{5.20}$$

5.5 Discretization of Ordinary Derivatives of Distributed Order

Using the matrix approach, the discretization of a derivative of distributed order is very easy. Let us discretize the interval $[a, b]$, in which the order α is changing, using the grid with the steps $\Delta\alpha_k$, not necessarily equidistant. Then we have

$$_0D_t^{w(\alpha)}f(t) = \int_{\gamma_1}^{\gamma_2} w(\alpha) \, _0D_t^\alpha f(t) \, d\alpha \approx \sum_{k=1}^{p} w(\alpha_k)\Big(_0D_t^{\alpha_k}f(t)\Big)\Delta\alpha_k \quad (5.21)$$

$$\approx \sum_{k=1}^{p} w(\alpha_k)\Big(B_n^{\alpha_k}f_n\Big)\Delta\alpha_k = \Big(\sum_{k=1}^{p} B_n^{\alpha_k}\, w(\alpha_k)\, \Delta\alpha_k\Big)f_n. \quad (5.22)$$

In other words, the discrete analog of distributed-order differentiation is given by the matrix that we will further denote as $B_{n,p}^{w(\alpha)}$,

$$B_{n,p}^{w(\alpha)} = \sum_{k=1}^{p} B_n^{\alpha_k}\, w(\alpha_k)\, \Delta\alpha_k, \quad (5.23)$$

and we can obtain the values of the distributed-order derivative at all points t_j $(j = 1, \ldots, n)$ at once using the following relationship:

$$_0D_t^{w(\alpha)}f(t) \approx B_{n,p}^{w(\alpha)}f_n. \quad (5.24)$$

In the notation $B_{n,p}^{w(\alpha)}$ the order $w(\alpha)$ means the function describing the distribution of orders α in the interval $[\gamma_1, \gamma_2]$, and the second index p is the number of discretization steps for α; the first index n, as above, is the number of discretization steps with respect to the variable t.

The visualization of the formula (5.21) is shown in Fig. 5.1. On each k-th layer of the shown "cake" the input vector f_n of the values of the function $f(t)$ at the nodes t_j is multiplied by the matrix $B_n^{\alpha_k}$ and gives the output vector of the values of the fractional derivative $_0D_t^{\alpha_k}$ at the same nodes t_j. Those vectors $_0D_t^{\alpha_k}$ are then multiplied by weights $w(\alpha_k)$ and discretization steps $\Delta\alpha_k$, and the final summation with respect to k ("summation across the layers of orders") gives the vector of the distributed-order derivative $_0D_t^{w(\alpha)}$ evaluated at the nodes t_j $(j = 1, \ldots, n)$.

5.6 Discretization of Partial Derivatives of Distributed Order

In contrast with generally used numerical methods, where the solution is obtained step-by-step by moving from the previous time layer to the next one, let us consider the whole time interval of interest at once. This allows us to create a net of

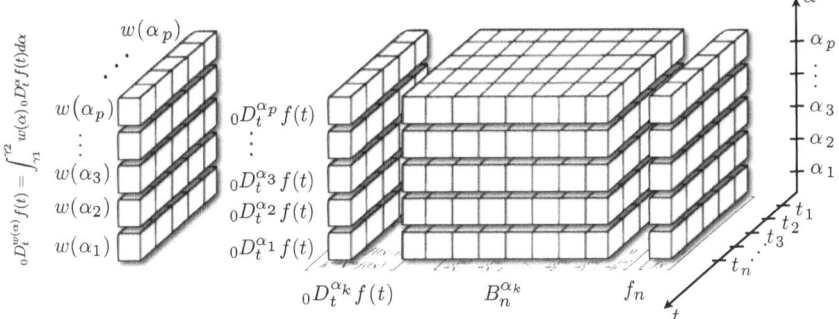

Fig. 5.1 Visualization of numerical evaluation of distributed-order derivatives with the help of the matrix approach

discretization nodes. In the simplest case of one spatial dimension this step gives a 2D net of nodes. An example of such discretization is shown in Fig. 5.2. The values of the unknown function in inner nodes (shaded area in Fig. 5.2) are to be found. The values at the boundaries are known, they are used later in constructing the system of algebraic equations.

The system of algebraic equations is obtained by approximating the equation in all inner nodes simultaneously (this gives the left-hand side of the resulting system of algebraic equations) and then utilizing the initial and boundary conditions (the values of which appears in the right-hand side of the resulting system).

The discretization nodes in Fig. 5.2 are numbered from right to left in each time level, and the time levels are numbered from bottom to top. We use such numbering for the clarity of presentation of our approach, although standard numberings work equally well.

The simplest implicit discretization scheme used for numerical solution of partial differential equations, like diffusion equation, is shown in Fig. 5.3, where the two nodes in time direction are used for approximating the time derivative, and the three points in spatial direction are used for the symmetric approximation of the spatial derivative. The stencil in Fig. 5.3 involves therefore only two time layers. If we consider fractional-order time derivative or distributed-order derivative, then we have to involve all time levels starting from the very beginning. This is shown in Fig. 5.4 for the case of five time layers.

Similarly, if in addition to time derivative of distributed order or of fractional order we also consider symmetric spatial derivatives of distributed order or fractional order, then we have to use all nodes at the considered time layer. This most general situation is shown in Fig. 5.5.

Let us consider the nodes $(ih, j\tau)$, $j = 0, 1, 2, \ldots, n$, corresponding to all time layers at i-th spatial discretization node. Similarly to the case of constant fractional orders (Podlubny 2000), all values of α-th order time derivative of $u(x, t)$ at these nodes are approximated using the discrete analogue of distributed-order differentiation:

Fig. 5.2 Nodes and their right-to-left, and bottom-to-top numbering

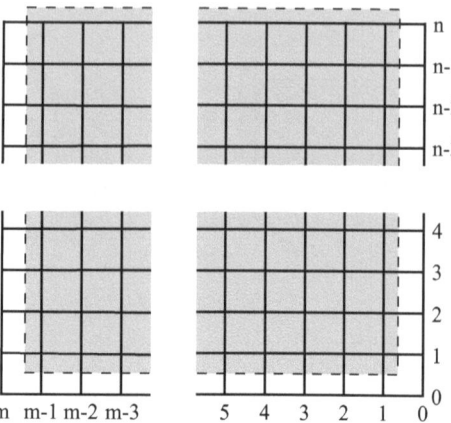

$$\left[u_{i,n}^{(w(\alpha))} \ u_{i,n-1}^{(w(\alpha))} \ \cdots \ u_{i,2}^{(w(\alpha))} \ u_{i,1}^{(w(\alpha))} \ u_{i,0}^{(w(\alpha))} \right]$$

$$= B_{n,p}^{w(\alpha)} \left[u_{i,n} \ u_{i,n-1} \ \cdots \ u_{i,2} \ u_{i,1} \ u_{i,0} \right]^T. \tag{5.25}$$

In order to obtain a simultaneous approximation of α-th order time derivative of $u(x, t)$ in all nodes shown in Fig. 5.2, we need to arrange all function values u_{ij} at the discretization nodes to the form of a column vector:

$$U_{nm} = \Big[u_{m,n} \ u_{m-1,n} \ \cdots \ u_{1,n} \ u_{0,n}$$

$$u_{m,n-1} \ u_{m-1,n-1} \ \cdots \ u_{1,n-1} \ u_{0,n-1}$$

$$\cdots \cdots \cdots$$

$$u_{m,1} \ u_{m-1,1} \ \cdots \ u_{1,1} \ u_{0,1}$$

$$u_{m,0} \ u_{m-1,0} \ \cdots \ u_{1,0} \ u_{0,0} \Big]^T. \tag{5.26}$$

In visual terms of Fig. 5.2, we first take the nodes of n-th time layer, then the nodes of $(n-1)$-th time layer, and so forth, and put them in this order in a vertical column stack.

The matrix that transforms the vector U_{nm} to the vector $U_i^{w(\alpha)}$ of the partial derivative of distributed order $w(\alpha)$ with respect to time variable can be obtained as a Kronecker product of the matrix $B_{n,p}^{w(\alpha)}$, which corresponds to the ordinary derivative of distributed order $w(\alpha)$ (recall that n is the number of time steps), and the unit matrix E_m (recall that m is the number of spatial discretization steps):

$$T_{mn}^{w(\alpha)} = B_{n,p}^{w(\alpha)} \otimes E_m. \tag{5.27}$$

This is illustrated in Fig. 5.6, where the nodes denoted as white and grey are used to approximate the fractional-order time derivative at the node shown in grey.

Similarly, the matrix that transforms the vector U to the vector $U_x^{\varphi(\beta)}$ of the derivative of distributed order $\varphi(\beta)$ with respect to spatial variable can be obtained as a Kronecker product of the unit matrix E_n (recall that n is the number of spatial discretization nodes), and the matrix $R_{m,p}^{\varphi(\beta)}$, which corresponds to symmetric Riesz ordinary derivative of distributed order $\varphi(\beta)$ (recall that m is the number of time steps):

$$S_{mn}^{\varphi(\beta)} = E_n \otimes R_{n,p}^{\varphi(\beta)}. \tag{5.28}$$

This is also illustrated in Fig. 5.6, where the nodes denoted as black and grey are used to approximate the symmetric fractional-order Riesz derivative at the same node shown in grey.

Having these approximations for partial fractional derivatives with respect to both variables, we can immediately discretize, for example, the diffusion equation in terms of time- and space-derivatives of distributed order by simply replacing the derivatives with their discrete analogs (Fig. 5.7). Namely, the equation

$$_0D_t^{w(\alpha)} u - \chi \frac{\partial^{\varphi(\beta)} u}{\partial |x|^{\varphi(\beta)}} = f(x, t) \tag{5.29}$$

is discretized as

$$\left\{ B_{n,p}^{w(\alpha)} \otimes E_m - \chi E_n \otimes R_{m,p}^{\varphi(\beta)} \right\} u_{nm} = f_{nm}. \tag{5.30}$$

5.7 Initial and Boundary Conditions for Using the Matrix Approach

It is always emphasized in case of the matrix approach to solution of differential equations that initial and boundary conditions must be equal to zero. If it is not so, then an auxiliary unknown function must be introduced, which satisfies the zero initial and boundary conditions. In this way, the non-zero initial and boundary conditions moves to the right-hand side of the equation for the new unknown function. After obtaining the solution for the auxiliary function, the backward substitution gives the solution of the original equation.

5.8 Implementation in MATLAB

A set of MATLAB routines implementing the described method is provided for download (Podlubny 2011). Those routines require the previously published toolbox

Fig. 5.3 A stencil for integer-order derivatives

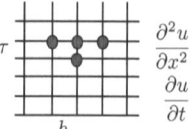

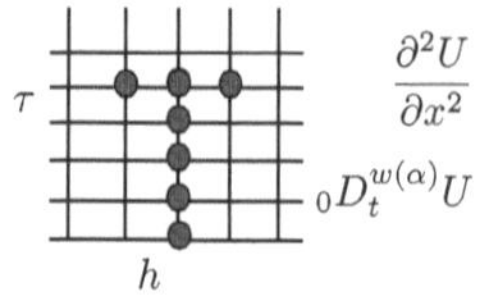

Fig. 5.4 A stencil in case of distributed-order and fractional time derivative and second-order spatial derivative

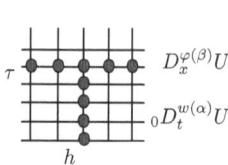

Fig. 5.5 A stencil in case of distributed-order and fractional-order time and spatial derivatives

Fig. 5.6 Discretization of partial derivaives

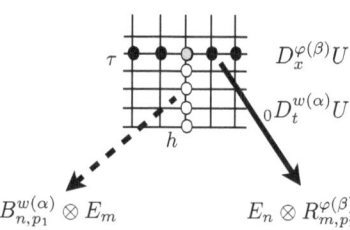

for numerical solution of differential equations of arbitrary (fractional) constant order (Podlubny et al. 2008).

The function DOBAN returns the matrix for the backward difference approximation of the left-sided distributed-order derivative, the function DOFAN returns the matrix for approximating the right-sided distributed-order derivative, DORANORT return the matrix for approximating the symmetric Riesz distributed-order derivative.

The use of these routines is illustrated by the demo functions included in the toolbox.

5.9 Numerical Examples

The use of the matrix approach for numerical solution of differential equations with derivatives of distributed orders is illustrated below on three examples that generalize the standard frequently used models of the applied fractional calculus. The relaxation, oscillation, and diffusion equations play extremely important role in numerous fields

Fig. 5.7 Discretization of partial derivaives and of the equation

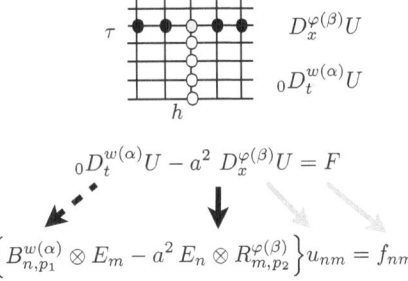

$$_0D_t^{w(\alpha)}U - a^2\, D_x^{\varphi(\beta)}U = F$$

$$\left\{ B_{n,p_1}^{w(\alpha)} \otimes E_m - a^2\, E_n \otimes R_{m,p_2}^{\varphi(\beta)} \right\} u_{nm} = f_{nm}$$

of science and engineering, and, because of their importance, they are also often used for benchmarking new methods and algorithms.

To the knowledge of the authors, these are the first examples of numerical solution of such distributed-order problems.

It is worth mentioning that existence and uniqueness of solutions of such types of distributed-order differential equations in the particular case of $w(\alpha) = 1$ were investigated by Pskhu (2005).

5.9.1 Example 1: Distributed-Order Relaxation

Let us consider the following initial value problem for the distributed-order relaxation equation:

$$_0D_t^{w(\alpha)}x(t) + bx(t) = f(t), \tag{5.31}$$

$$x(0) = 1, \tag{5.32}$$

where the distribution of the orders α is given by the function $w(\alpha) = 6\alpha(1-\alpha)$, $(0 \le \alpha \le 1)$. To be able to use the matrix approach, we need zero initial condition. Introducing an auxiliary function $u(t)$,

$$x(t) = u(t) + 1$$

gives the following initial value problem for the new unknown $u(t)$:

$$_0D_t^{w(\alpha)}u(t) + bu(t) = f(t) - b, \tag{5.33}$$

$$u(t) = 0. \tag{5.34}$$

The discretization of equation (5.33) gives the following system of algebraic equations in the matrix form:

Fig. 5.8 Solution of the distributed-order relaxation equation with $w(\alpha) = 6\alpha(1 - \alpha)$

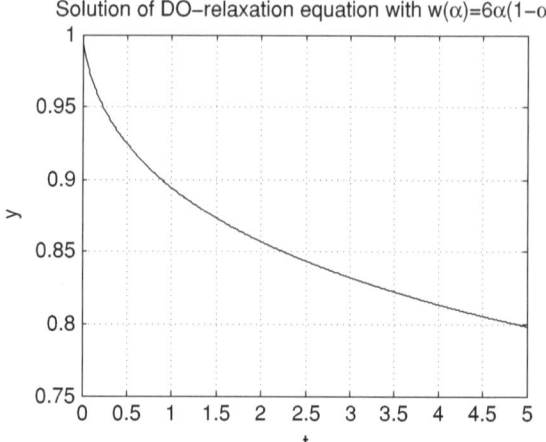

Solution of DO–relaxation equation with w(α)=6α(1–α)

$$\left(B^{w(\alpha)}_{n,p} + bE_n\right)U_n = F_n, \tag{5.35}$$

where U_n is the vector of the values of $u(t)$ at the discretization nodes, and F_n is the vector of the values of the right-hand side, $f(t) - B$, at the same nodes; E_n is the identity matrix. The MATLAB code for solving Example 1 is in the Appendix, and the results of computations are shown in Fig. 5.8.

5.9.2 Example 2: Distributed-Order Oscillator

Let us consider the Bagley-Torvik equation with a damping term described by a distributed-order derivative. When the damping term is of constant (integer or non-integer) order, this equation is also called the fractional oscillator equation.

$$ay''(t) + by^{(w(\alpha))}(t) + cy(t) = f(t), \qquad f(t) = \begin{cases} 8, & (0 \le t \le 1) \\ 0, & (t > 1) \end{cases}, \tag{5.36}$$

$$y(0) = y'(0) = 0. \tag{5.37}$$

Similarly to Example 1, we just replace continuous operators with their corresponding discrete analogs in the form of matrices, and the known and unknown function by the vectors of their values in the discretization nodes. This gives the following algebraic system in the matrix form:

$$\left(a B^2_n + b B^{w(\alpha)}_{n,p} + c\right)Y_n = F_n. \tag{5.38}$$

Fig. 5.9 Solution of the distributed-order oscillator equation (Bagley-Torvik equation) with $w(\alpha) = 6\alpha(1 - \alpha)$

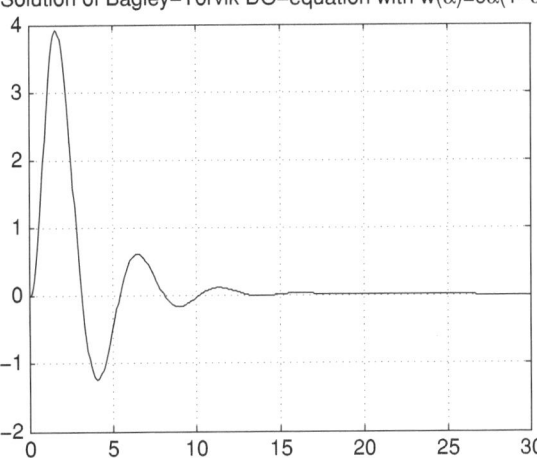

Solution of Bagley–Torvik DO–equation with w(α)=6α(1−α)

The MATLAB code for solving Example 2 is in the Appendix, and the results of computations are shown in Fig. 5.9.

5.9.3 Example 3: Distributed-Order Diffusion

The last example that is provided in this section is an initial value problem for a partial differential equation with a derivative of distributed-order $w(\alpha)$ with respect to time variable t and with a constant-order symmetric fractional derivative of order β with respect to the spatial variable x:

$$_0D_t^{w(\alpha)}y - \frac{\partial^\beta y}{\partial|x|^\beta} = f(x, t) \tag{5.39}$$

$$y(0, t) = 0, \quad y(1, t) = 0; \qquad y(x, 0) = 0. \tag{5.40}$$

In order to be able to check for a "backward compatibility" of the obtained solution with the solution of the classical diffusion equation and with the solution of the constant-order fractional diffusion equation, we take $f(x, t) = 8$.

Again, we just replace continuous operators with their corresponding discrete analogs in the form of matrices, and the known and unknown function by the vectors of their values in the discretization nodes. This gives the following algebraic system in the matrix form:

$$\left(B_{n,p}^{w(\alpha)} \otimes E_m - E_n \otimes R_m^\beta\right)Y_{nm} = F_{nm}. \tag{5.41}$$

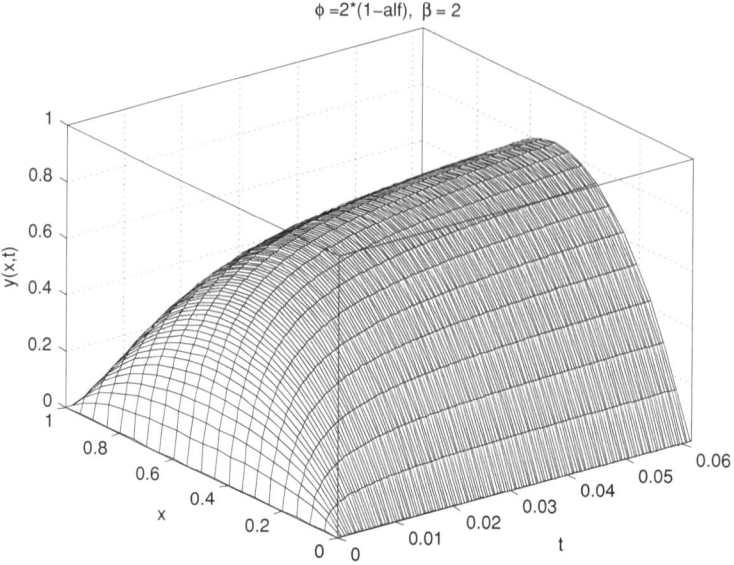

Fig. 5.10 Solution of distributed-order diffusion equation with $w_1(\alpha) = 2(1 - \alpha)$

Here we demonstrate that distributed-order derivatives, integer-order derivatives and fractional-order derivatives can appear in the same equations and are treated in the same manner for the purpose of numerical solution using the matrix approach.

The MATLAB code for solving Example 3 is also provided in the Appendix. The results of computations are shown in Fig. 5.10 for the case $w_1(\alpha) = 2(1 - \alpha)$, and in Fig. 5.11 for the case $w_2(\alpha) = 2\alpha$. Although we have

$$\int_0^1 w_1(\alpha)\, d\alpha = \int_0^1 w_2(\alpha)\, d\alpha = 1,$$

the obtained solutions are different because of different weights assigned by the functions $w_1(\alpha)$ and $w_2(\alpha)$ to the fractional derivatives of orders close to 0 and 1.

5.10 Chapter Summary

In this chapter we introduced the extension of Podlubny's matrix approach to the case of distributed-order derivatives. The matrix approach provides an extremely convenient language and framework for discretization of differentiation of any order—integer, fractional, and distributed order. Using discrete analogs of all those forms of differentiation, one can easily discretize differential equations with all

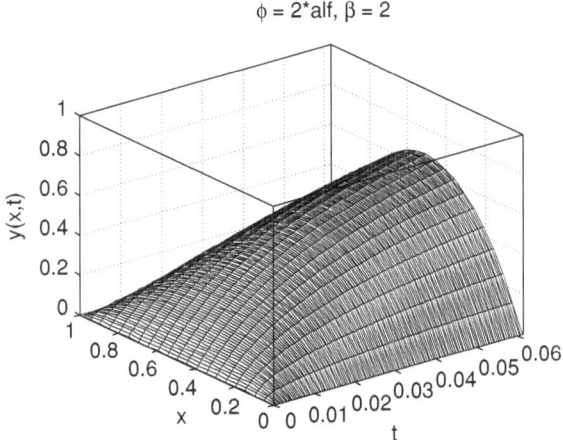

$\phi = 2*\mathrm{alf}, \beta = 2$

Fig. 5.11 Solution of distributed-order diffusion equation with $w_2(\alpha) = 2\alpha$

possible combinations of derivatives—classical integer-order derivatives, left- and right-sided fractional order derivatives, symmetric fractional derivatives, and left-sided, right-sided, and symmetric distributed-order derivatives.

We have provided examples of solution of the three important types of problems that are important for applications and appear in many fields of science and engineering. The MATLAB code provided in the Appendix demonstrates how easy it is using the matrix approach.

References

Bulgakov BV (1954) Kolebaniya (Vibrations). Gostekhizdat, Moscow

Gantmakher FR (1988) Theory of matrices. Nauka, Moscow

Loan CFV (2000) The ubiquitous Kronecker product. J Comput Appl Math 123:85–100

Ortigueira MD (2006) Riesz potential operators and inverses via fractional centred derivatives. Int J Math Math Sci 48391:1–12

Ortigueira MD, Batista AG (2008) On the relation between the fractional Brownian motion and the fractional derivatives. Phys Lett A 372:958–968

Podlubny I (2000) Matrix approach to discrete fractional calculus. Fract Cal Appl Anal 3(4):359–386

Podlubny I (2011) Matrix approach to distributed-order ODEs and PDEs. http://www.mathworks.com/matlabcentral/fileexchange/

Podlubny I, Skovranek T, Vinagre BM (2008) Matrix approach to discretization of ODEs and PDEs of arbitrary real order. http://www.mathworks.com/matlabcentral/fileexchange/22071

Podlubny I, Skovranek T, Vinagre BM (2009a) Matrix approach to discretization of ordinary and partial differential equations of arbitrary real order: the matlab toolbox. In: Proceedings of the ASME 2009 international design engineering technical conferences and Computers and information in engineering conference IDETC/CIE 2009 August 30–September 2, 2009, San Diego, USA, article DETC2009-86944

Podlubny I, Chechkin A, Skovranek T, Chen YQ, Vinarge BM (2009b) Matrix approach to discrete
    fractional calculus ii: partial fractional differential equations. J Comput Phys 228(8):3137–3153
Podlubny I, Skovranek T, Verbickij V, Vinagre BM, Chen YQ, Petras I (2011) Discrete fractional
    calculus: non-equidistant grids and variable step length. In: Proceedings of the ASME 2011
    international design engineering technical conferences and computers and information in
    engineering conference IDETC/CIE 2011 August 28–31, 2011, Washington, DC, USA, article
    DETC2011-47623
Pskhu AV (2005) Differencial'nye uravneniya drobnogo poryadka (Partial differential equations of
    fractional order). Nauka, Moscow (in Russian)
Skovranek T, Verbickij V, Tarte Y, Podlubny I (2010) Discretization of fractional-order operators
    and fractional differential equations on a non-equidistant mesh. In: Proceedings of the 4TH IFAC
    workshop on fractional differentiation and its applications, Badajoz, Spain, October 18–20, 2010,
    article FDA10-157
Suprunenko DA, Tyshkevich RI (1966) Commutative matrices. Nauka i Tekhnika, Minsk

Chapter 6
Future Topics

In previous chapters, methods and tools for the modeling of distributed-order systems were discussed, which include stability analysis of distributed-order systems in four cases of the weighting function of order, and two special cases: double noncommensurate orders and N-term noncommensurate orders. Distributed-order signal processing technique and optimal distributed-order damping strategies were studied. A general approach to numerical solution to discretization of distributed-order derivatives and integrals, and to numerical solution of ordinary and partial differential equations of distributed order was proposed.

Let us now outline possible topics for future research in the field of distributed-order systems.

6.1 Geometric Interpretation of Distributed-Order Differentiation as a Framework for Modeling

Distributed-order derivatives can be considered as a kind of averaging operators. However, they do not average the values of parameters over a set of values—they average the behavior of individual objects or processes over a set of objects or processes of the same type.

Let us consider an understandable example of the impact of advertising on the crowd consisting of a large number of individuals. The broadcast of an advertisement is a unit-step input; this can be interpreted as charging the memory of individuals to some initial value.

The impact of such input on an individual vanishes in time due to properties of human's memory (memory relaxation, or forgetting). Suppose it is possible to classify individuals with respect to the properties of their memory into p groups. The process of memory relaxaation in the i-th group of individuals can be described by an initial-value problem for a two-term fractional differential equation in terms of Caputo derivatives with non-zero initial condition:

Z. Jiao et al., *Distributed-Order Dynamic Systems,* SpringerBriefs in Control, Automation and Robotics, DOI: 10.1007/978-1-4471-2852-6_6,
© The Author(s) 2012

$$_0\mathrm{D}_t^{\alpha_i} x(t) + bx(t) = 0, \qquad x(0) = 1, \tag{6.1}$$

where the order α_i is from the interval $[0, 1]$; we assume that $\alpha_1 < \alpha_2 < \ldots < \alpha_p$.

Denoting w_i the number of individuals in i-th group normalized with respect to the total number of individuals (so that $\sum w_i = 1$), we obtain a piecewise-constant weighting function

$$w(\alpha) = w_i \quad \text{for } \alpha_i < \alpha \le \alpha_{i+1} \quad (i = 1, \ldots, p-1). \tag{6.2}$$

Then the averaged behavior of the crowd can be described as

$$\sum_{i=1}^{p-1} w(\alpha_i) \, _0\mathrm{D}_t^{\alpha_i} x(t) \, \Delta\alpha_i = \, _0\mathrm{D}_t^{w(\alpha)} x(t), \tag{6.3}$$

which is a distributed-order derivative, and the equation describing the process of the memory relaxation of a crowd takes on the form of a distributed-order differential equation

$$_0\mathrm{D}_t^{w(\alpha)} x(t) + bx(t) = 0, \tag{6.4}$$

$$x(0) = 1. \tag{6.5}$$

Of course, in many situations the function $w(\alpha)$ describing the distribution of orders α can be modeled as a continuous function.

The outlined general framework for creation of distributed order models can be used for modeling multiscale and multifractal processes, multiagent systems, and in many other cases where the subject of study consists of "individuals".

The above considerations are presented in graphical form in Fig. 6.1, which gives a geometric interpretation of distributed-order differentiation and integration.

6.2 From Positive Linear Time-Invariant Systems to Generalized Distributed-Order Systems

Let us consider a linear differential equation with fractional-order derivatives,

$$\sum_{k=1}^{n} a_k \, _0\mathrm{D}_t^{\alpha_k} x(t) = f(t), \tag{6.6}$$

where all coefficients are positive constants: $a_k > 0, k = 1, \ldots, n$. Without the loss of generality, we can assume that

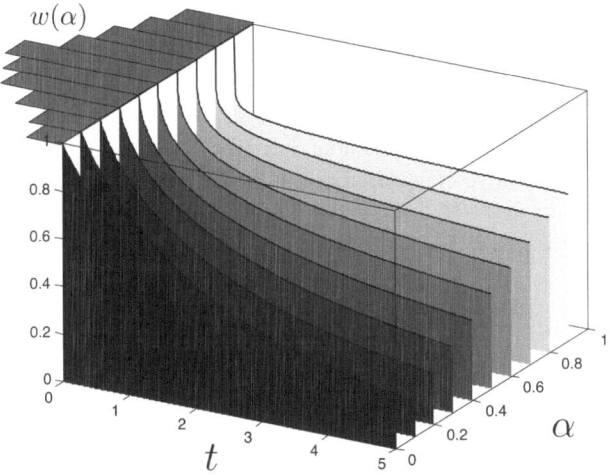

Fig. 6.1 Geometric interpretation of distributed-order differentiation as a framework for creating distributed-order models

$$\sum_{k=1}^{n} a_k = 1.$$

Introducing the weighting function

$$w(\alpha) = \sum_{k=1}^{n} a_k \delta(\alpha - \alpha_k), \tag{6.7}$$

we can write the Eq. (6.6) in the form of a distributed-order differential equation:

$${}_0D_t^{w(\alpha)} x(t) = f(t). \tag{6.8}$$

This means that properties of positive linear time-invariant systems should follow from the properties of distributed-order systems with a positive function $w(\alpha)$ describing the distribution of orders.

Such a change of the viewpoint means also that the problem of identification of parameters of linear time invariant systems like (6.6) can be replaced by a more general and more interesting problem of identification of a most appropriate shape of the function $w(\alpha)$ of a system described by Eq. (6.7).

Similarly, the theory of stability of linear time-invariant systems can be overcome by the theory of stability of distributed-order systems.

Even more possibilities open if the distributed-order derivative is written in the form

$$0\tilde{D}_t^{w(\alpha)}x(t) = \int_{\gamma_1}^{\gamma_2} {}_0D_t^{\alpha}x(t)\mathrm{d}W(\alpha). \tag{6.9}$$

If the function $W(\alpha)$ is differentiable and $W'(\alpha) = w(\alpha)$, then we have the distributed-order derivative considered in this book; otherwise there is large space for further generalizations with potential applications to self-similar, fractal and multifractal systems.

6.3 From PID Controllers to Distributed-Order PID Controllers

PID Controller

The classical *PID* controller, which is the most commonly used feedback controller in industry, can be considered as a particular form of lead-lag compensation in the frequency domain. Its transfer function can be expressed as $C_1(s) = K_p + \frac{K_i}{s} + K_d s$. The *PID* controller algorithm involves three separate constant parameters: the proportional P, the integral I and derivative D. These values can be interpreted in terms of time: P depends on the present error, I on the accumulation of past errors, and D is a prediction of future errors, based on current rate of change.

Fractional-Order $PI^{\lambda}D^{\mu}$ Controller

Podlubny (1999) proposed a generalization of the *PID* controller, namely the $PI^{\lambda}D^{\mu}$ controller. Its transfer function can be expressed as $C_2(s) = K_p + \frac{K_i}{s^{\lambda}} + K_d s^{\mu}$, which involves an integrator of order λ and a differentiator of order λ. Better response of this type of controller in comparison with the classical *PID* controller was demonstrated in that work. For results about using $PI^{\lambda}D^{\mu}$ controller to some real world systems, please refer to Li et al. (2009) and Luo and Chen (2009) etc.

Distributed-Order $PI^{\lambda(\alpha)}D^{\mu(\alpha)}$ Controller

As a generalization of $PI^{\lambda}D^{\mu}$ controller, distributed-order *PID* controller of the following form

$$C_3(s) = K_p + \int_{-1}^{1} w(\alpha)s^{\alpha}\mathrm{d}\alpha$$

where $w(\alpha) = \sum_{i=0}^{N} w_i \alpha^i \approx w_0 + w_1 \alpha$

or

$$C_3(s) = K_p + K_i \int_0^{\gamma_1} \frac{1}{s^\alpha} \mathrm{d}\alpha + K_d \int_0^{\gamma_2} s^\alpha \mathrm{d}\alpha$$

will be our future work.

References

Li HS, Luo Y, Chen YQ (2009) A fractional order proportional and derivative (FOPD) motion controller: tuning rule and experiments. IEEE Trans Control Syst Technol 18(2):1–5

Luo Y, Chen YQ (2009) Fractional order [proportional derivative] controller for a class of fractional order systems. Automatica 45(10):2446–2450

Podlubny I (1999) Fractional-order systems and $PI^\lambda D^\mu$ controllers. IEEE Trans Autom Control 44(1):208–214

Appendix A
MATLAB Codes

A.1 Stable Boundary of Distributed-Order Linear Time-Invariant Systems

```
% The following matlab code describes how to plot the stable
% boundary of distributed-order linear time-invariant
% system
            x(1) = 0;      % set the initial value of x
            y(1) = 0;      % set the initial value of y

% the computed number of the stable boundary
w = logspace(-200, 2, 10000);

% the stable boundary of distributed-order
% system of case 1: W (\alpha) = 1
for i = 1 : length (w)
        x(i) = (2*pi*w(i) - 4*log(w(i)))/(4*(log(w(i)))^2 + pi^2);
        y(i) = (4*w(i)*log(w(i)) + 2*pi)/(4*(log(w(i)))^2 + pi^2);
end

% the stable boundary of distributed-order
% system of case 2: w(\alpha) = \alpha
for i = 1 : length(w)
        x(i) = ((-1 - log(w(i)))*((log(w(i)))^2 - pi^2/4) + ...
        pi*log(w(i))*(w(i) - pi/2))/((log(w(i)))^2 + pi^2/4)^2;
        y(i) = ((w(i) - pi/2)*((log(w(i)))^2 - pi^2/4) + ...
        pi*log(w(i))*(1 + log(w(i))))/((log(w(i)))^2 + pi^2/4)^2;
end

% the stable boundary of distributed-order
% system of case 3: w(\alpha) = \delta(\alpha - \beta)
beta = 0.5;
```

Z. Jiao et al., *Distributed-Order Dynamic Systems*, SpringerBriefs in Control, Automation and Robotics, DOI: 10.1007/978-1-4471-2852-6,
© The Author(s) 2012

```
for i = 1 : length(w)
    x(i) = w(i)^beta*cos(pi*beta/2);
    y(i) = w(i)^beta*sin(pi*beta/2);
end

% the stable boundary of fractional-order
% system with double noncommensurate orders
beta1 = sqrt (2) − 1;
beta2 = sqrt (3) − 1;
for i = 1 : length(w)
        x(i) = w(i)^beta1*cos(pi*beta1/2) + w(i)^beta2*cos(pi*beta2/2);
        y(i) = w(i)^beta1*sin(pi*beta1/2) + w(i)^beta2*sin(pi*beta2/2);
end

% plot the stable boundary
plot(x,y)
hold on
plot(x,-y)
axis square
grid on
xlabel('Real axis');
ylabel('Imag axis');
```

A.2 Bode Plots of Distributed-Order Linear Time-Invariant Systems

```
% The following matlab code describes how to plot the Bode
% plots of distributed-order linear time-invariant system

function [bd] = bode_dos(Ts)

% Ts: The sampling period
if Ts < = 0,
sprintf('%s','Sampling period has to be positive'),
return,
end

% rad./sec. Nyquist frequency
wmax0 = 2*pi/Ts/2;

wmax = floor(log10(wmax0)) + 1;
wmin = wmax − 5;
w = logspace(wmin,wmax,1000);
j = sqrt(−1);

% the Bode plot of distributed-order
% system of case 1: w(\alpha)=1
srfr = log(j.*w)./(j.*w − 1 + log(j.*w));
```

```
% the Bode plot of distributed-order
% system of case 2: w(alpha)=\alpha
srfr = (log(j.*w)).^2./(j.*w - 1 - log(j.*w) + log(j.*w).*log(j.*w));

% the Bode plot of distributed-order
% system of case 3: w(\alpha)=\delta(\alpha-\beta)
srfr = 1./((j.*w).^0.5 + 1);

% the Bode plot of fractional-order
% system with double noncommensurate orders
%srfr = 1./((j.*w).^(sqrt(3) - 1) + (j.*w).  ^0.5 + (j.*w).^(sqrt(2) - 1) + 1);

% plot the magnitude property
subplot(2,1,1);
semilogx(w,20*log10(abs(srfr)));
hold on;
ylabel('Magnitude (dB)');
grid on;

% plot the phase property
subplot(2,1,2);
semilogx(w,(180/pi)*(angle(srfr)));
xlabel('Frequency (Hz)');
ylabel('Phase (degrees)');

end
```

A.3 Impulse Responses of Distributed-Order Linear Time-Invariant Systems

```
% The following matlab code describes how to plot the impulse responses
% of distributed-order linear time-invariant system based on NILT
% technique
% Code adapted from

% L. Bran.cL.k. Programs for fast numerical inversion of Laplace
% transforms in MATLAB language environment. In Proceedings of the
% 7th Conference MATLABf99, pages 27.39, Prague, Nov. 1999. Czech
% Republic.

function [id] = BICO_irid(Ts);

% Ts: The sampling period
% the impulse response of distributed-order
% system of cases 1: w(\alpha)=1
F = C*log(s).*((s - 1)*eye(2) - A*log(s)).^(-1)*B;

% the impulse response of distributed-order
% system of cases 2: w(\alpha) = \alpha
F = C*log(s).^2.*((s - 1 - log(s))*eye(2) - A*log(s).^2).^(-1)*B;
```

```
% the impulse response of distributed-order
% system of case 3: w(\alpha)=\delta(\alpha-\beta)
F = C*(s.^beta*eye(2) − A).^(−1)*B;

% the Bode plot of fractional-order
% system with double noncommensurate orders
F = C*((s.^beta1 + s.^,beta2)*eye(2) − A).^(−1)*B;

alfa = 0;
M = 3072;
P = 20;
Er = 1e − 10;
tm = M*Ts;
wmax0 = 2*pi/Ts/2;
L = M;
Taxis = [0 : L − 1]*Ts;
n=1:L-1;
n = n*Ts;
q1 = q;
N = 2*M;
qd = 2*P + 1;
t = linspace(0, tm, M);
NT = 2*tm*N/(N − 2);
omega = 2*pi/NT;
c = alfa − log(Er)/NT;
s = c − i*omega*(0 : N + qd − 1);
Fsc = feval(F, s);
ft = fft(Fsc(1 : N));
ft = ft(1 : M);
q = Fsc(N + 2 : N + qd)./Fsc(N + 1 : N + qd − 1);
d = zeros(1, qd);
e = d;
d(1) = Fsc(N + 1);
d(2) = −q(1);
z = exp(−i*omega*t);

for r = 2 : 2 : qd − 1
    w = qd − r;
    e(1 : w) = q(2 : w + 1) − q(1 : w) + e(2 : w + 1);
    d(r + 1) = e(1);
    if r > 2
    q(1 : w − 1) = q(2 : w).*e(2 : w)./e(1 : w − 1);
    d(r) = −q(1);
    end
end
A2 = zeros(1, M);
B2 = ones(1, M);
A1 = d(1)*B2;
B1 = B2;
for n = 2 : qd
A = A1 + d(n)*z.*A2;
B = B1 + d(n)*z.*B2;
```

```
A2 = A1;
B2 = B1;
A1 = A;
B1 = B;
end

ht = exp(c*t)/NT.*(2*real(ft + A./B) − Fsc(1));
[b1, a1] = stmcb(ht(1 : end).*Ts, q1, q1, 100);

sprintf('Impulse response of approximated transfer function:')

sr = tf(b1, a1, Ts);
hhat = impulse(sr, Taxis);
figure
plot(t, ht, 'b');
grid on;

xlabel('time axis');
ylabel('impulse response');
xlim([0,30])
```

A.4 Solution of Distributed-Order Relaxation Equation

```
% (1) Prepare constants and nodes
% (this is the longest part of the script):
h = 0.01;                  % step of discretization in space
t = 0:h:5;                 % step of discretization in time
N = length(t) + 1;         % number of nodes
B = 0.1;                   % coefficient of the equation
f = '0 + 0*t';             % right-hand side
M = zeros(N, N);           % pre-allocate matrix M for the system

% (2) Write the discretization matrix
M = doban('6*alf.*(1 − alf)', [0 1], 0.01, N − 1, h) + B*eye(N − 1, N − 1);

% (3) Compute the right-hand side at discretization nodes:
F = eval ([f' − B'], t)';

% (4) Utilize zero initial condition:
M = eliminator(N − 1, [1])*M*eliminator(N − 1, [1])';
F = eliminator(N − 1, [1])*F;

% (5) And solve the system MY = F :
Y = M\F;

% (6) Pre-pend the zero initial value
% (that one due to zero initial condition)
Y0 = [0; Y];
```

```
% Plot the solution:
U = Y0 + 1;
plot(t, U, 'k')
```

A.5 Solution of Distributed-Order Oscillation Equation

```
% (1) Prepare constants and nodes
% (this is the longest part of the script):
A = 1; B = 1; C = 1;      % coefficients of the Bagley-Torvik equation
h = 0.075;                % step of discretization
T = 0:h:30;               % nodes
N = 30/h + 1;             % number of nodes
M = zeros(N, N);          % pre-allocate matrix M for the system

% (2) Make the matrix for the entire equation --- this is really easy:
M = A*ban(2, N, h) + B*doban('6*alf.*(1 − alf)', [0 1], 0.01, N, h) + ...C*eye(N, N);

% (3) Make right-hand side:
F = 8*(T < = 1)';

% (4) Utilize zero initial conditions:
M = eliminator(N, [1 2])*M*eliminator(N, [1 2])';
F = eliminator(N, [1 2])*F;

% (5) Solve the system MY = F :
Y = M\F;

% (6) Pre-pend the zero values (those due to zero initial conditions)
Y0 = [0; 0; Y];

% Plot the solution:
plot(T, Y0, 'k')
grid on
```

A.6 Solution of Distributed-Order Diffusion Equation

```
alpha =' 2*alf'  ; beta = 2;   % First, define the orders:
a2 = 1;      % coefficient from the diffusion equation
L = 1;       % length of spatial interval

% Number of spatial steps + 1 is:
m = 21;% 11, 21
% Number of steps in time + 1 is:
n = 148;% 37, 148
h = L/(m − 1); tau = h^2/(6*a2);   % space step, time step
```

```
% generating the matrix for approximation
% alpha-th order derivative with respect to time
B1 = doban(alpha, [0 1], 0.01, n − 1, tau)′;
TD = kron(B1, eye(m));          % time derivative matrix
B2 = ransym(beta, m, h);        % beta-th order derivative with respect to X
SD = kron(eye(n − 1), B2);      % spatial derivative matrix
SystemMatrix = TD − a2∗SD;      % matrix corresponding to
                                       discretization
                                % in space and time


% remove columns with '1' and 'm' from SystemMatrix
S = eliminator (m, [1 m]); SK = kron(eye(n − 1), S);
SystemMatrix_without_columns_1_m = SystemMatrix∗SK′;

%   remove  rows  with  '1'  and  'm'  from  SystemMatrix_without_
    columns_1_m
S = eliminator(m, [1 m]); SK = kron(eye(n − 1), S);
SystemMatrix_without_rows_columns_1_m = ...
                    SK∗SystemMatrix_without_columns_1_m;
% Right hand side
F = 8∗ones(size(SystemMatrix_without_rows_columns_1_m, 1), 1);

% Solution of the system
Y = SystemMatrix_without_rows_columns_1_m\F


% Reshape solution array -- values for k-th time step
% are in the k-th column of YS:
 YS = reshape(Y, m − 2, n − 1);
 YS = fliplr(YS);
 U = YS;


% plot graph
[rows, columns] = size(U);
U = [zeros(1, columns); U; zeros(1, columns)];
U = [zeros(1, m)′  U];
[XX, YY] = meshgrid(tau∗(0 : n − 1), h∗(0 : m − 1));

mesh(XX, YY, U)
xlabel('t'); ylabel('x'); zlabel('y(x, t)');
title(['\phi =', num2str(alpha), ', \beta = ', ... num2str(beta)])
set(gca, 'xlim', [0 tau*n], 'zlim', [0  1]); box on
```

Index

B
Bounded-input bounded-output stability, 25

C
Caputo's fractional-order derivative
    definition, 3
Constant phase element, 4
CRONE control, 5

D
Differentiator, 45
Digital filter, 43
Distributed-order damping, 48
Distributed-order differential equation, 6
Distributed-order differential/integral opera-
    tors, 5
Distributed-order diffusion equation, 7
Distributed-order diffusion-wave equation, 7
Distributed-order fractional mass-spring
    viscoelastic damper, 51, 53
Distributed-order integrator/differentiator, 8,
    19, 39, 40, 42, 55
Distributed-order linear time-invariant system,
    8
Distributed-order low-pass filter, 7, 45, 55
Distributed-order partial differential equation,
    7
Distributed-order signal processing, 8, 55
Distributed-order systems, 6
Distributed-order wave equation, 7
Distributed-parameter systems, 2
Double noncommensurate orders, 30

F
Fractional calculus operators, 5
Fractional calculus, 2
Fractional filter, 39
Fractional-order controller, 5, 48, 49
Fractional-order differential equation, 3
Fractional-order diffusion equation, 4
Fractional-order diffusion-wave equation, 5
Fractional-order integral equation, 3
Fractional-order partial differential
    equations, 4
Fractional-order system, 5, 7
Fractional-order wave equation, 5
Function of distribution of order, 6

G
Generating series, 60
Grünwald-Letnikov's fractional-order
    derivative/integral definition, 2

H
Heat equation, 2
Human's memory, 75

I
Impulse response invariant
    discretization, 43, 47
Integer-order controller, 5
Integer-order system, 5, 7
Integral of absolute error, 49
Integral of squared error, 49, 53

Z. Jiao et al., *Distributed-Order Dynamic Systems*, SpringerBriefs in Control,
Automation and Robotics, DOI: 10.1007/978-1-4471-2852-6,
© The Author(s) 2012

I (*cont.*)
Integral of time multiplied absolute
    error, 49
Integral of time multiplied squared
    error, 49

K
Kronecker product, 61

L
Lower triangular strip matrices, 60
Lumped-parameter systems, 2

M
Mass-spring viscoelastic damper, 50
Multiscale and multifractal processes, 76
Multi-valued function, 12

N
Noncommensurate orders, 8
N-term noncommensurate orders, 33
Numerical inverse Laplace transform, 34

O
Optimal fractional-order damping, 53
Ordinary and partial differential equations of
    distributed order, 8
Ordinary differential equation, 1

P
Partial differential equation, 1
$PI^\lambda D^\mu$ controller, 5
PID controller, 78

R
Realization of fractional system, 43, 47
Riemann-Liouville's fractional-order
    derivative definition, 3
Riemann-Liouville's fractional-order integral
    definition, 3

S
Self-similar, fractal and multifractal
    systems, 78
Support set, 7
Symmetric Riesz derivative, 63

T
Triangular strip matrices, 59

U
Upper triangular strip
    matrices, 60

W
Warburg impedances, 4
Wave equation, 2